Using the Telescope

Using the Telescope

A Handbook for Astronomers

J. Hedley Robinson

David & Charles

Newton Abbot   London   North Pomfret (VT)   Vancouver

British Library Cataloguing in Publication Data
Robinson, J Hedley
    Using the telescope.
    1. Telescope—Amateur's manuals
    I. Title
    522'.2   QB88
    ISBN 0-7153-7592-X

Set by Ronset Ltd, Darwen, Lancashire
and Printed in Great Britain
by Biddles Limited, Guildford
for David & Charles (Publishers) Limited
Brunel House Newton Abbot Devon

Published in the United States of America
by David & Charles Inc
North Pomfret Vermont 05053 USA

Published in Canada
by Douglas David & Charles Limited
1875 Welch Street North Vancouver BC

Contents

Foreword   6

Introduction   7

Instruments   9

Time, the Clock and Position   32

Observing the Sun   37

General Introduction to Lunar and Planetary Observation   46

Observing the Moon   47

Observing Mercury   57

Observing Venus   61

Observing Mars   67

Observing the Asteroids   73

Observing Jupiter   75

Observing Saturn   80

Observing Uranus, Neptune and Pluto   87

Observing Comets and Meteors   89

Observing the Stars   93

Observing Star Clusters, Nebulae and Galaxies   103

Brief Glossary   105

Appendices

1 Magnitudes of Asteroids   106

2 Longitude on Saturn   106

3 Magnitudes of Comets   107

4 Stellar Magnitudes   107

5 National Bodies   109

Bibliography   110

Index   111

Foreword

After constructing or purchasing a suitable telescope the would-be observer often finds that, apart from pleasurable viewing of the celestial bodies, he is at a loss as to what he can do.

The author of this little manual here offers the results of some fifty years of experience at the telescope in the hope that it will be of help to observers, especially the beginners.

Thanks are gratefully extended to those who have willingly assisted by their permission to include illustrations, and whose names appear under the appropriate pictures.

Introduction

Astronomy has a double appeal: its study reveals the grandeur and design of the Universe, and gives the assiduous observer the opportunity to take part in work that can be of scientific value. At the same time, one must remember that a large area of the science can be explored only by professionals, with the backing of sophisticated apparatus and public funds. On the other hand there are areas in which the private or amateur observer can do observational work of real value.

It is to these people, who are willing to work assiduously, enduring the hardship of cold winter nights, sacrificing social pleasures and spending their own money on apparatus, that I offer my admiration and support. It is for the benefit of such heroic souls that this little manual has been written in the hope that one man's experience might not only encourage them, but be of practical help.

Firstly, I would offer some suggestions on the practical matter of keeping warm in the night air: wear loose clothing and shoes. Pressure on the body restricts the flow of blood and cold ensues. Wear mittens which leave the fingertips free to handle papers, etc., and not gloves, which impede the fingers. Felt slippers help to keep the feet warm, especially if the soles are thick.

Secondly, I would suggest that there are many fields of observation open to the telescope user: the Sun, the Moon, the planets, comets, stars and nebulae. These are all treated in the appropriate sections that follow. Before proceeding to these, it would be as well for the reader to consult the brief glossary on page 105 in order to understand the most common technical terms used in observational astronomy.

Instruments

Choice of Instrument

Since telescopes vary, not only in type but from one specimen to another, it is difficult to lay down exact recommendations or instructions for any particular instrument. However, the reader will be able to apply the general principles outlined in the following pages to his particular instrument.

Wide-angle views at low magnification are often required, and binoculars supply this need, but for the serious planetary or lunar observer they are of very limited use. Binoculars provide an extension of naked-eye viewing, but do not satisfy the need for high magnification, for which a telescope is a necessity.

Among small instruments the refractor holds pride of place, being far superior to the small reflector of, say, less than 6in (150mm) diameter. As size increases, the reflector overtakes the refractor, until in large instruments the reflector is supreme, mainly because of loss of light in passage through the relatively thick objective lens of the refractor. Under poor conditions the small refractor can supply steadier images than the reflector of similar power. The advantages and disadvantages of the two types are set out below.

REFLECTORS

Advantages	*Disadvantages*
Colour-free images	Temperature changes affect adjustment
Cheapness in relation to diameter	More maintenance necessary
Shorter tube than similar refractor	Scatter of light by 'spider' support of secondary mirror can be detrimental to the image
Easier observing position with the Newtonian type	
Good light grasp in large sizes	

9

REFRACTORS

Advantages	Disadvantages
Useful in sizes from 3in (75mm) upwards	Colour effects can be annoying
	Cost is higher in relation to size
Little adjustment required	Longer tube can be a nuisance
Maintenance virtually nil	Mounting must be high for
Closed tube minimises tube currents	observation near the zenith
No refracting obstructions (such as the 'spider' in a reflector)	(All the disadvantages increase rapidly with size).

The matter of size is governed by the type of observation to be undertaken. Table 1 illustrates this.

Table 1

Diameter of objective lens or mirror		Limiting magnitude	Limit of resolution (seconds of arc)
3in	(76mm)	9·9	1·8
6in	(152mm)	11·6	0·9
8in	(203mm)	12·3	0·6
10in	(254mm)	12·8	0·5
15in	(380mm)	13·8	0·3

These figures presume the mirror of a reflector is in good condition.

Now if you know the limiting magnitude you wish to reach, say for observation of variable stars or other faint objects, and the resolution required for lunar or planetary work, you will be able to choose a suitable size of instrument.

When buying a telescope, or even if you are making one with optical parts you have bought, you must check its performance on a star. To view terrestrial objects, such as a church clock some miles away, is not enough.

Rack out the eyepiece until the image of the star is out of focus and check the shape of the diffraction rings produced by the light of the star under high magnification. These should be circular. Any distortion may be due to pinching of the objective lens or mirror in its cell, or to an optical defect. Astigmatism may sometimes be cured by rotating the lens or mirror in its cell. The distorting effect can also be produced by undue pressure on the flat in its mounting on a Newtonian or similar type of reflector.

Colour should be minimal in a good refractor, but will still be present even in the best of telescopes. If the image shows a red halo around the star when in focus, the lens is under-corrected. Over-corrected lenses produce a blue halo, but this is better than a red one as the eye does not pick up the blue so well as the red. In any case, if the colour is not pro-

1 A Newtonian reflector on an altazimuth mounting (*A. W. Heath*)

nounced, you probably have a good lens, especially if the colour of the halo is bluish.

With all this in mind, decide on the size of instrument you intend to choose and its type, bearing in mind that your telescope is a precision instrument and will require proper mounting. Furthermore, if it is too large to be easily portable, it will require weatherproof housing.

Mountings

Mountings for astronomical telescopes fall into two classes. The first is the altazimuth, with movement in altitude and azimuth (or, in common language, movement in the vertical and horizontal planes), and the second is the equatorial type, with movement in Right Ascension and declination.

Altazimuth

For terrestrial use the altazimuth mounting (Plate 1) is very satisfactory, but when one comes to observing celestial objects, which rise in the east, reach their culmination on the meridian, and then set in the west, one is trying to follow a movement that is in quite different planes from those in which the altazimuth mounting moves. Furthermore, celestial objects appear to travel in a curve from east to west, reaching their highest altitude on the meridian; thus to follow a celestial object on an altazimuth mounting it is necessary to move the instrument in both altitude and azimuth at the same time.

Equatorial

This problem is overcome by tilting the vertical axis until it is parallel with the Earth's axis of rotation, in which situation the horizontal movement becomes parallel with the equator (both being regarded as produced onto the apparent celestial sphere). Movement is now in Right Ascension and declination, and the apparent movement of the celestial object under observation is reduced to one plane, in which the telescope can be made to move; that is, in Right Ascension or Hour Angle, of which both are on the same plane.

Equatorial mountings can be classified into several types.

The German Type, which is most common, has the telescope on one side of the mounting polar axis and counterbalanced by a weight on the other side, on an extension of the declination axis. One disadvantage of this type is the fact that it is necessary to reverse the instrument when the object passes the meridian.

Normally the telescope is used above the counterbalance weight so that it clears any obstruction by the mounting itself, and is thus easier to use than when the telescope is below the counterbalance weight. On the other hand, the German type is very suitable for the medium-sized refractor and is a compact arrangement.

The Fork or Yoke Type has the telescope swinging between the prongs of the fork, and movement in Right Ascension is by rotation of the fork on its long axis. This type is often preferred, especially for large reflectors. If sufficient clearance is arranged, the mirror of a reflector can be made to clear the inside of the fork or yoke, and the observer has access to the eyepiece at the upper end of the Newtonian type of instrument (Plate 2). In the case of the refractor it is impossible to get access to the eyepiece when the instrument is pointed to the pole on this type of mounting. The Cassegrain type of reflector obviously suffers the same disadvantage as the refractor. It is, of course, not necessary to reverse the instrument mounted in a fork or yoke.

The English Type has the telescope swinging between two parallel beams aligned along the polar axis of the instrument and supported at both ends by separate piers. As a variant of this type, the telescope may be mounted on one side of a single beam or similar axis (a lorry back-axle has been used most satisfactorily), and counterbalanced by a weight on the other side of the axis. The great drawback in this type is the space required for its erection and operation, in addition to which with the double-beams version, the pole cannot be reached.

The Coudé Type has two mirrors, with the telescope mounted above

2 A Newtonian reflector on an equatorial mounting of the fork type (*D. Sinden*)

them and forming the polar axis. Movement of the first mirror provides adjustment in declination, and rotation of the telescope on its longer axis produces movement in Right Ascension. Rotation of the field results from the movement in Right Ascension, and can be confusing.

There are many developments of these types, but the principles of movement in Right Ascension and declination remain the same.

Rigidity is the prime consideration in all types, coupled with easy movement, and it is therefore usual to make the mounting heavy to avoid vibration, especially when mechanical automatic drive is fitted. Mountings for refractors must be high enough to enable the observer to get below the instrument to observe the zenith; but reflector mountings should be as low as possible, so that the observer does not have to climb high ladders to get at the eyepiece, except in the case of the Cassegrain telescope, where the eyepiece is behind the main mirror.

Housing

This is necessary for anything larger than the small portable instruments. The simplest form of housing is the run-off shed, which is virtually an inverted box with doors at one end to enable it to be run off from the instrument on wheels, sometimes on rails. The next simplest is the run-off roof, which may be in two or more sections for ease of handling. Various methods can be devised to support the roof when it is run off for observing.

Finally we consider the rotating observatory or rotating dome on a fixed building (Plate 3). This must have an adequate opening for observation and may rotate on wheels, rollers or balls.

In any of the types mentioned it is necessary for the telescope to clear the building in all positions, and extra space should be provided for the observer to move round the instrument when observing with it. Construction of any of the types mentioned is not beyond the capabilities of the average handy-man, but he must think carefully when planning.

Eyepieces

It can be argued that an ordinary magnifying glass may be used as an eyepiece (or ocular, as it is sometimes called) for a telescope. Should you not be critical of the quality of the image, this may be agreed, but if you require proper rendering with a flat field (i.e., all parts of the field of view brought to a focus at the same plane) and good images, then you will need something better.

Single lens eyepieces, such as the Tolles type, are used, but these are in a very different class from the common magnifying glass. Other than the Tolles type, eyepieces are generally compound; that is, they are made up of more than one lens. Ordinarily there is a field lens system, with an

3 Left, the author's 11ft 6in (3·5m) dome revolving on golf balls and made of marine plywood with hardboard cladding; and right, rectangular shed with run-off roof covered in PVC sheeting.

eye lens or lenses for examination of the image formed by the field lens and the objective or mirror.

The main types of eyepiece (Fig 1) are listed below.

Huyghenian eyepieces, which are most common for refractors, have two planoconvex lenses separated by a distance of two-thirds of the focal length of the field lens, both having the plane surfaces towards the eye. The image falls between the two lenses, making them unsuitable for use with a cross wire or occulting bar, although such devices have been mounted between the lenses at the critical focus of the eye lens with some success. They are not suitable for short-focal-length reflectors.

Ramsden types of eyepieces are often used for reading scales, etc., and are composed of two planoconvex lenses mounted with the convex sides facing each other. The image falls outside the system, and the relative focal length is the product of the several focal lengths divided by their sum minus the distance between them. The two lenses are of similar focal lengths, and the Ramsden type can be used on both refractors and reflectors.

Kellner or achromatic Ramsden systems have a large field lens and wide angle of view, but have the disadvantage of the focal plane being on the face of the field lens. Thus, every speck of dust is magnified and seen in the field of view. However, they serve well for variable stars, clusters, comets and nebulae where large fields with bright images are necessary.

Tolles eyepieces are formed of one piece of glass, and in this respect are similar to the magnifying glass mentioned above. They are excellent for short-focus reflectors, since they work down to f/7 very well. The main drawback is the small field of view, but for planets and lunar detail they are good.

Orthoscopic types are made up of a triple colour-corrected field lens and a viewing lens, and have a large field of view. They are excellent for planetary and lunar work. They are reputed to be ghost-free—that is, free from internal reflections—but this point should be checked before buying, since I have come across examples with ghosts.

Monocentric eyepieces in the achromatic form are excellent for planetary work and work well with reflectors. In the latter case it is sometimes well worth while fitting a Barlow lens to flatten the field.

Erfle types are sometimes used for wide fields at low magnification, but opinion seems to be divided as to their individual performance. It would be as well to try one out with your particular instrument before buying it.

It is a common fallacy to think magnification should be pushed to the theoretical limit of the instrument. Clarity of vision is the ultimate aim, and this depends on seeing conditions as well as the quality of the objective

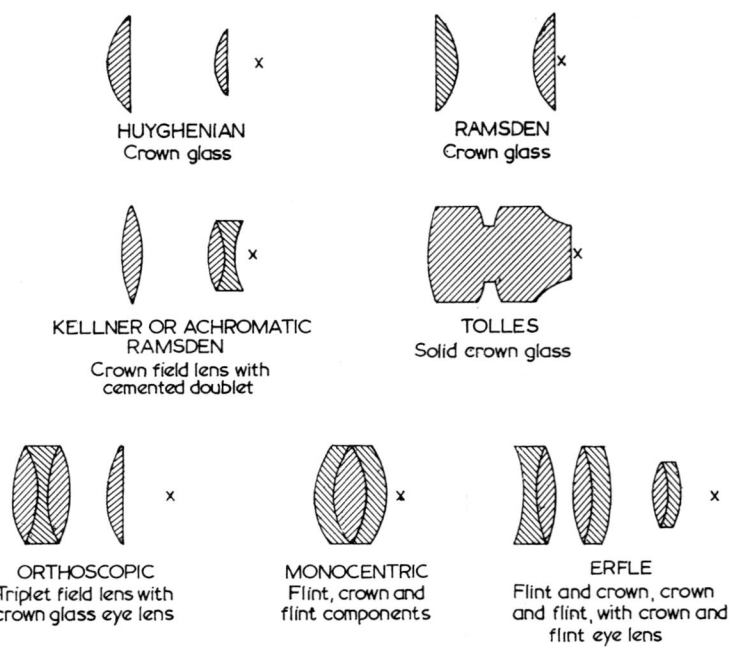

HUYGHENIAN
Crown glass

RAMSDEN
Crown glass

KELLNER OR ACHROMATIC
RAMSDEN
Crown field lens with
cemented doublet

TOLLES
Solid crown glass

ORTHOSCOPIC
Triplet field lens with
crown glass eye lens

MONOCENTRIC
Flint, crown and
flint components

ERFLE
Flint and crown, crown
and flint, with crown and
flint eye lens

x - eye position

Fig 1 Eyepieces. Crown glass lenses are hatched to the left, and flint glass lenses to the right

or mirror. One hundred times magnification per inch of objective diameter is often stated as the limit. This may on occasion be so for small refractors, but for general use with larger instruments fifty times per inch is nearer the practical mean, while for large telescopes this figure drops to thirty times per inch diameter. Metrically, a good guide is twice the diameter of the telescope in millimetres.

At the other extreme, wide fields and low magnification are required for comets and nebulae. Here it is well to limit the size of the exit pupil so that all light available finds its way into your eye and is not spilled over your face! A lower limit of three times magnification per inch of objective or mirror is about right.

The main faults in eyepieces are the following:

1 Spherical aberration, causing flare off axis, usually bad in low powers;
2 Chromatic aberration, producing changes in colour with position, and objects outlined in red against a light sky, blue in a dark sky;
3 Distortion, which shows a straight line as convex to the centre at the edge of the field;

4 Field curvature, producing lack of uniform sharpness;
5 Astigmatism, rendering a pattern sharp at one angle and indistinct at other angles; and
6 Chromatic difference, which causes colour-free images at the centre with the field edges coloured.

A battery of eyepieces should include one of low power followed by a series of, say, four or five with gradually decreasing focal lengths. Do not go in for very short-focus eyepieces, because you will not use them. In operation you will choose by trial and error the best magnification to use for observation under the conditions prevailing at the time. The focal length will depend on the size and type of instrument being used and the local conditions. Generally speaking, focal lengths of eyepieces fall between about 2·5in (65mm) and, at the lowest extreme, about 0·2in (5mm).

Adjusting the Equatorial Mounting

Much has been written on the use of the telescope and its care by amateur astronomers, but comparatively little appears on the subject of the proper setting of the mounting of an equatorial instrument. Correct setting is necessary if the instrument is to give satisfaction and to enable an observer to locate and follow accurately any celestial object under observation.

It is not sufficient to align the polar axis with the Pole Star and correct by the error in following. While this may suffice for small instruments and occasional viewing of bright objects like the Moon, etc., it does not enable one to locate with any degree of accuracy an object not directly visible in the finder telescope or other sighting devices.

In the following remarks it is presumed the telescope is satisfactory in other respects, and especially that the optical axis is accurately set at right-angles to the declination axis. It remains to set the polar axis parallel to the Earth's axis of rotation. The basic rule to remember is that the altitude of the pole always equals the latitude of the place of observation.

It will be seen by reference to Fig 2 that at the equator observer A sees the pole on his horizon with altitude zero, while at the pole observer B sees the pole directly overhead with altitude 90°. Thus the altitude of the pole equals the latitude of the observer. Intermediate places follow the same pattern, as observer C at latitude 52° sees the pole at that altitude.

With the altitude of the polar axis approximately set, the final adjustment of the zero reading of the declination circle is accomplished by taking readings of the declination of any star near the meridian, with the telescope both sides of the mounting. In the case of the fork mounting

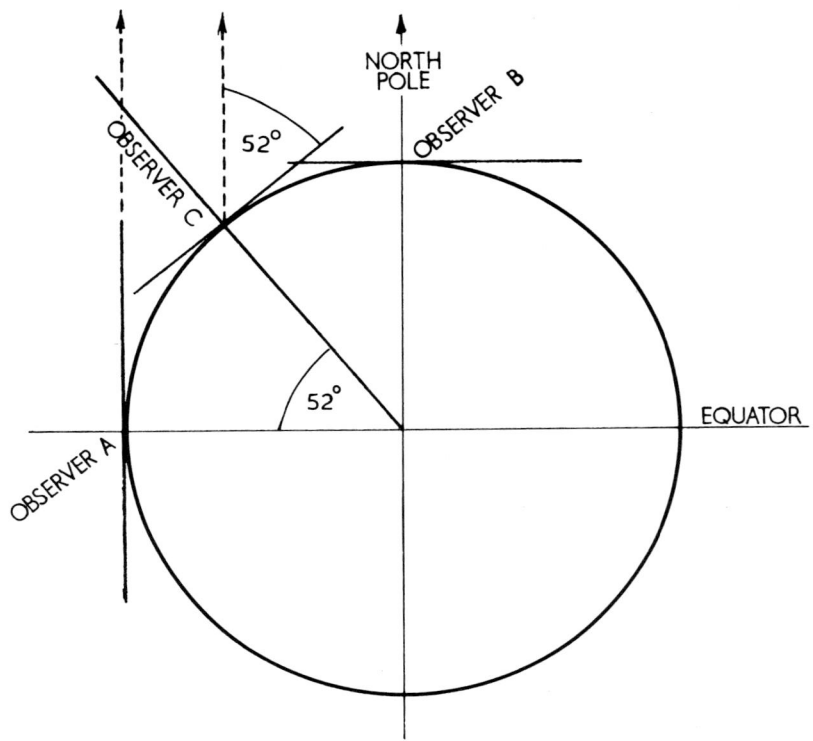

Fig 2 The altitude of the pole always equals the latitude of the observer

only one reading is required, but for the more normal German type the telescope is set alternately east and west of the mounting, and both readings of the declination of the star are recorded. The vernier of the declination circle is adjusted and the two readings east and west are again taken. The vernier is set by repeated observations and adjustments in ever-decreasing values until the correct reading is reached both sides of the mounting. Should any appreciable error persist, the polar axis may be found to be out of alignment; it can be adjusted to take it up, in which case the polar axis should be moved slightly in altitude only, and the observations repeated until the setting is correct. Any error in the gradua-tion of the declination circle will now become apparent, but can generally be allowed for in ordinary use of the telescope.

Now select two stars of similar declination, one east and the other west of the meridian. Set the eastern star in the field of view of the telescope and read the apparent declination. Reverse the instrument and repeat, using the western star with the telescope east of the mounting. Record both readings in declination.

If the western star is too far south, then the polar axis lies in a NE-SW direction from the meridian. If the western star is too far north, then the axis lies NW–SE. Slew the whole mounting slightly in azimuth accordingly and repeat both observations and record both the amount of slew and the resultant declination readings. Repeat this operation until both stars east and west of the meridian read correctly on the declination circle. Check by again observing the declination of the meridian star.

If all is well, your instrument is now in its correct position, with the polar axis accurately in line with the Earth's axis, and the mounting is correctly set in azimuth. Now take from your star catalogue the declination of any star and set the declination circle to read the correct value. Point the instrument in the direction of the star without altering the declination setting, and sweep. The star should appear in the field of view with a low-power eyepiece.

Repeat this type of observation in various parts of the sky as a final check. You should have little trouble in picking up any object you desire to observe, even if it is not visible in the finder telescope.

Further, when making prolonged observations of stars, nebulae, or planets, your instrument should follow without your altering the declination setting. With the Moon, its motion north or south in its orbit may necessitate some small adjustment if you are using a large magnification.

The adjustments described above may seem somewhat tedious, but are thoroughly worth while once they are done. They will add considerably to the efficiency of your instrument.

Astronomical Photography

Astronomical photography falls under three main headings, as follows:
1 Wide-angle pictures of comets, star clusters or parts of constellations, and nebulae;
2 Higher magnification for photographs of the Moon and planets; and
3 Solar photography.

Wide-angle Pictures

These pictures, of comets and star fields, etc., are obtained by using short-focal-length cameras (f/3 is quite satisfactory for amateur work) attached to a driven telescope for accurate guiding during the comparatively long exposure time necessary to obtain images of the fainter stars. Good pictures can be obtained using portrait lenses and masking off the outer parts of the field where distortion may occur. The guiding telescope should be of focal length at least three times that of the camera.

Sometimes an ordinary camera may be strapped to the tube of an equatorially mounted telescope and time exposures made while the telescope tracks the stars as they move with the Earth's diurnal motion.

It is well to choose a fairly bright star for tracking and to put the image slightly out of focus in the high-power eyepiece. This should be fitted with cross wires or some similar device, and the star kept steadily in the field of view. It is easier to keep a star central when out of focus on a cross wire than to attempt to keep a clearly focused star behind a wire. To follow the motion, the telescope should be equatorially mounted, and small adjustments taken up as the exposure progresses, both in Right Ascension and in declination. Few mountings are accurate enough to be left alone to do the following by themselves, and atmospheric refraction will alter the rate of movement as the horizon is approached. For this kind of work films or plates should be fairly rapid, to catch as many faint stars as possible in the minimum of time (Plates 4 and 5).

Long exposures can produce poor images through fatigue in visual guiding and atmospheric conditions causing wandering of the image. Film fogging can also reduce the quality of the picture if too lengthy exposures are made. A little trial and error will establish the best length of exposure to suit the lens, emulsion and sky conditions.

Colour film, which can now be obtained in fairly fast ranges, produces most attractive pictures, but it must be remembered that the colours are dependent on film response and are not those seen by the human eye. Kodak Ektachrome-X and Anscochrome films have been used with success.

4 The Hyades and Pleiades photographed with 2in (50mm) Tessar lens and 1 min exposure on Tri-X (*A. W. Heath*)

5 The Southern Cross and Coalsack photographed from Hobart, Tasmania, with 3·5in (90mm) lens at f/1 and 10 min exposure, showing stars to the eleventh magnitude (*W. E. Pennell*)

Lunar and Planetary Photography
Such photography demands high contrast to bring out detail (Plate 6), and it is best to limit the exposure time to the minimum to avoid trouble with following the object to keep it in the same place on the film continuously. Photographs may be taken with an ordinary camera, focused to infinity, held close to the eyepiece, which must be critically focused beforehand. Better results are generally obtained by disposing of the camera lens and using an eyepiece to enlarge the prime image on the film. Careful arrangement is necessary if the image is to be focused accurately.

The photographic limiting magnitude follows the rule that

$$Mp = \frac{5 \log D + 2 \cdot 15 \log E + 6}{M^2 d^2}$$

where Mp is the photographic limiting magnitude,
  D is the diameter of the objective or mirror in inches,
  E is the length of exposure in minutes,
  M is the magnification used,
  d is the distance of the film or plate from the eyepiece, also in inches.
Sometimes a negative lens, such as the Barlow, is used for this purpose and produces excellent pictures. It must, however, be remembered that the brightness of the image falls off in proportion to the square of the

magnification employed. Prime-focus images may also be used, but considerable enlargement is necessary to produce the final picture.

With eyepiece projection the effective focal length of the system is

$$E=\frac{LF}{af}$$

where E is the effective focal length,
L is the distance from the focal plane to the eyepiece,
F is the focal length of the mirror or objective lens,
a is the aperture of the mirror or objective lens,
f is the focal length of the eyepiece.

Small-grain film is recommended to retain the detail, yet the exposure should be kept to the shortest time possible. Faster films have larger grains.

The focal length of the optical system determines the size of the image (Table 2). It will be seen that for planets' smaller subtended angles the images are very small in comparison.

Table 2

Focal length (in)	Diameter of Moon's image (in)
45	0·4
60	0·54
78	0·7

While photographs of the Moon and planets supply a definite record that is accurate as to detail, it must be understood that photography of these bodies will not show as much in the way of fine detail as a drawing made at the eyepiece of a telescope by a practised observer. This is a fact of life and not the fault of the photographer. (Compare Plates 7 with 16 and 8 with 17).

The apparent diameters of the various planets are mentioned in the appropriate sections of this book. They can be compared with the apparent diameter of the Moon, which is about half a degree.

Solar Photography
This should never be attempted at the prime focus, for the heat will burn the emulsion and the gelatine film. A neutral filter preferably over the objective lens, or an unsilvered mirror or diagonal is advised. Exposure should be as short as possible, say 1/100 to 1/500 sec. The size of the solar

6 Region of the Lunar Alps taken with a 12in (305mm) mirror at f/30 and exposure of 1 sec (*Cdr. H. R. Hatfield*)

image will approximate that of the Moon as above.

Alternatively, the usual projection of the sun on a screen can be utilised by photographing the screen in the normal manner. In this case the camera should be mounted as close to the optical axis as possible to avoid distortion of the disk in the final picture. Normal camera speeds are employed in this system and no danger to the camera can occur.

As a general rule it is true to say that the normal photographic practices apply to astronomical photography, but the most practical way is to use the trial and error method. Instruments and conditions are so extremely variable that to lay down definite rules is fraught with danger and disappointment. Take the usual precautions of assessing the length of exposure, etc., then make an exposure, recording full details, and see what the result is. Proceed from that point and adjust exposure, etc., accordingly until a satisfactory result is obtained. Printing on contrasty paper is an obvious course, and lengthened developing time can produce improvements, but still the experimental method is advisable.

Some guidance on the use of an ordinary camera may be called for. The Moon, for instance, is a Sun-lit object and will require exposures

7 Jupiter photographed with 12in (305mm) mirror at f/85 and exposure of 1 sec. Compare with Plate 16 (*Cdr. H. R. Hatfield*)

8 Saturn taken with a 12in
(305mm) mirror at f/60.
Compare with Plate 17
(*Cdr. H. R. Hatfield*)

similar to those used for terrestrial objects. Comets will require exposures of some 10-30 min. Aurorae will require the lens at its widest opening, a fast film, and exposures varying between 1 sec and 2 min. It is well to take a number of pictures at various exposures and then to choose the best specimen.

Star fields will need longer exposures, which, as we have already said, should be decided by trial and error. Remember that a star is a point source and its brightness depends on the diameter of the lens, while focal ratio affects the brightness of extended objects. As an example, one might note that an f/4 system will transmit four times as much light as an f/8 system, but for point sources the image intensity that reaches the film depends on the diameter of the objective of the camera lens or telescope and not on the 'f' number.

To sum up, one could say that astronomical photography is not an exact science but an art. The art lies in the correct balancing of exposure time; choice of film or plate speed and grain size; and choice of the best conditions for taking pictures, with the danger of sky fogging in long exposures. There is also an art in the selection of the development and printing processes.

We might repeat that the trial-and-error method is the best course for the would-be photographer to adopt.

Accessories

The telescope has been described as a light bucket, used to catch light. This may be true, but one must add the proviso that the light must be accurately focused in one plane. What you do with that light depends on the observational programme undertaken: you may need merely to examine the image with an eyepiece in the usual manner; to measure the small separation of two binary stars, or the arc subtended by some other small object; to measure the comparative brightness of two stars or other objects; or to discriminate between the images at various wavelengths.

For these purposes you will need, in addition to the light bucket, an eyepiece, a micrometer, a photometer and a set of filters.

Micrometers

Normally of the bi-filar type, micrometers have two parallel wires that can be moved to overlap or can be separated to give a measure of the distance between two points in the field of view. The wires are normally moved by a screw and the number of turns or fractions of turns of the screw recorded, and, after calibration with known objects, converted to seconds of arc. This entails the screw being accurately formed to avoid errors, and the cage holding the wires moving without any backlash on the screw.

To avoid these difficulties, I have devised a micrometer with a vertical wire and two others that slide across it. These two wires are inclined to each other at an angle of about one in ten, as in Fig 3. The screw is used only for moving the wires and need not be accurate, as the movement of the wires is read off on a scale in millimetres, and estimated under a magnifier to a tenth of a millimetre. As an example of the micrometer in use, let us take two stars whose separation is required. Place them first on the vertical wire; the moving wires are then moved across the stars until the two images are accommodated within the triangle formed by the three wires. The scale is read and the moving wires then moved to the position where they meet and no light passes through the triangle. This is the closure of the wires, and the scale is again read. The difference between the two positions on the scale is the amount of the separation in milli-metres on the scale. The instrument is calibrated by taking measures of binary stars whose separation is well known, and establishing the value of one millimetre on the scale in seconds of arc. The wires are mounted in the focal plane of an eyepiece of the positive type, and this eyepiece is always used. Changing the eyepiece results in changing the value of the scale divisions, as also does transfer of the micrometer to another tele-scope.

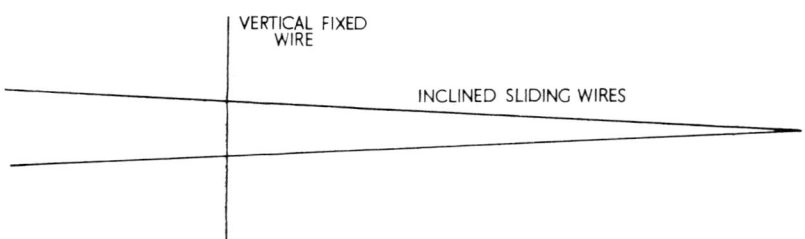

VERTICAL FIXED WIRE

INCLINED SLIDING WIRES

Fig 3 Micrometer with inclined moving wires

Another type of micrometer is the ring, but its use is governed by mathematical relationships between transits at different parts of a ring in the focal plane, and the full description is beyond the scope of this book. The more technical textbooks describe its use.

Photometers

Photometers for measuring the brightness of celestial objects are many and various in type and method of operation. A graduated wedge of tinted glass is sometimes employed to extinguish the star, but difficulty arises in deciding when a star is actually extinguished, since acuity of vision and weather conditions enter into the matter. An artificial star capable of variation of brightness may be employed, as in the Zollner type of photometer (see page 98). Here again difficulties arise through atmospheric conditions, and the photometer has to be recalibrated on each and every occasion it is used. For amateur work it has been found to be more reliable to adopt some method of comparing the brightness of a star with that of another star, or stars, of known brightness. Either the fractional or the step method may be employed, and full instructions in these methods are available through the Variable Star Section of the British Astronomical Association, and as set out on page 110.

Filters

These can never improve the quality of the image in a telescope, but they can help the observer to distinguish detail otherwise difficult to see. They are used to separate light of various wavelengths, and can clean up an image in the telescope, by suppressing unwanted light from blue to red. Orange or red filters will be found helpful for observing detail on Mercury or Mars; while, by suppressing the red, a blue filter will render detail on Jupiter easier to see. This applies especially to the cloud belts and the Red Spot.

A yellow filter is now accepted as standard for estimating the phases of Venus, while a pale blue helps to diminish the glare when observing the Moon. A neutral filter may also be used for this latter purpose.

Filters are normally used between the eyepiece and the eye, but can be placed between the two mirrors of a reflecting telescope. Placing a filter between the objective and eyepiece on a refractor necessitates cutting slots in the draw tube, with a consequent weakening of the tube, and is therefore not recommended.

Mounted between glass slips, gelatine filters are both cheap and satisfactory. They should be chosen with a view to their density and narrow cut. Do not choose one with high density, for it will extinguish the image; and a wide-cut filter will pass light over a large section of the spectrum and prove to be of little value. The following Kodak, Wratten types are suitable: 25 or 29 red, 15 yellow, and 47 or 44A blue.

Apodising Screen
Recently used in stellar and planetary work with success, this screen is composed of three similar layers of fine wire gauze, one layer having the wires horizontal and vertical, the second layer offset 30° to the right of the first layer, and the third offset 30° to the left. Holes are cut as percentages of the aperture of the instrument, as follows:

	1st layer (per cent)	2nd layer (per cent)	3rd layer (per cent)
Reflector	55	78	90
Refractor	52	76	88

The whole assembly is fitted over the open end of the reflector tube or over the objective of the refractor telescope. Diffraction spectra occur around the image but are not obtrusive, while the clarity of vision is greatly enhanced. Unfortunately the apodising screen cannot be used on extended objects like the Moon.

Maintenance of the Telescope

Refractors
The lenses forming the objective system should never be a tight fit in the cell. Just enough freedom should be allowed to enable the glass to expand with rising temperature, so that pressure does not cause distortion. With a small refractor, shake the cell and if you hear a rattle all is well. Do not remove the lenses from the cell unless absolutely necessary; if this has to be done, note the positions of the 'nicks' on the edges of the lenses, and see that they are replaced in exactly the same positions relative to each other.

Some telescopes are provided with screws for squaring-on the objective, but in others this is preset. In either case the objective system should be absolutely square with the tube. This can be checked by noting the shapes of the diffraction rings when a star is seen, with a medium magnification, out of focus. Any pressure points will be easily seen, and can be adjusted by repacking the lenses in the cell accordingly.

If the objective system is not truly squared-on with the tube, the diffraction rings of a star will appear oval, as they do if astigmatism is present in the lenses; these two effects can be distinguished by rotating the lens in the telescope tube.

To maintain the instrument in good working order it is necessary to take care of the objective lenses. On a damp night the dew can easily form on the face of the lens of the telescope, but this can largely be prevented by the use of a dew cap, which can be formed of a sheet of cardboard or metal, rolled to fit over the end of the telescope tube, and extending at least two and a half times the diameter of the objective lens. It is also a good plan to line the dew cap inside with blackened blotting paper; soaking the paper in ink and drying it out before use is a simple method of accomplishing this.

The objective should also have a tight-fitting cap to protect it from damage. This cap should always be kept on when the instrument is not in use. Remove the cap as a last measure before using the telescope, and replace it first when finishing. Never carry the instrument indoors and then place the cap on it, for the lens will dew over as soon as the instrument is carried into the warmer air inside. It is also a good idea to follow these rules with an instrument in an observatory. Never touch the objective lenses unless it is absolutely necessary. Cleaning should be carried out with extreme care so as not to scratch the glass or destroy the figure. First lightly brush off any dust with a soft-haired brush, holding the brush *under* the lens so that sharp particles are not ground into the glass but fall downwards away from it. If the lens is greasy, a little toilet soap and tepid water can be used. Do not flood the lens or water may find its way between the components, so leading to greater trouble and the possibility of having to take the lenses out of their cell. Pure alcohol will remove grease and help drying, and should be applied with a soft washed handkerchief or similar soft material. Never use new material or any form of abrasive.

Similar treatment may be given to the lenses of eyepieces. Do not rub the eyepiece lenses when observing, for in the dark you may easily cause damage with sharp grit on the surfaces of the glasses.

Reflectors

Most reflector telescopes are supplied with covers for both the main mirror and the flat, which should be kept covered when not in use. The general rule is to prepare the observatory and instrument, then take the cover off the flat. To avoid the danger of damage to the main mirror, the cover should be removed from it last of all. When observation is finished, the main mirror should be immediately covered. Sometimes the cover is lined with damp-absorbing material, which can be dried out during the observing period, but I have found with my 10in (260mm) mirror, in a wooden observatory constantly at the same temperature as the outside air, that no lining is necessary. The cover fits closely over the mirror and traps the minimum of air between itself and the mirror surface.

The tube of a reflector acts as a dew cap, and little trouble should be experienced with the main mirror. The best procedure with a portable instrument is to keep the mirror covered and only open it up immediately before using, covering it again immediately observation is finished and before carrying it indoors.

The flat may occasionally dew over. The warmth of a hand held over

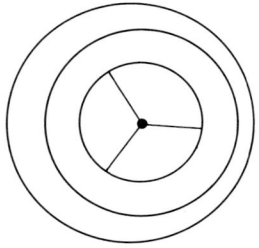 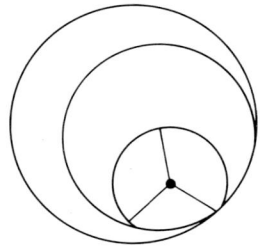

Fig 4 Mirrors on a Newtonian reflector: *left*, out of adjustment; *right*, adjusted

the mounting of the flat can often clear this away, but care should be taken not to touch the flat mirror itself. Some observers fit an electric torch bulb behind the flat mirror, which, when duly covered with foil, transmits a little heat to the flat without emitting any light.

A reflector will not perform properly unless it is correctly collimated; i.e., the optical centres of the mirrors should be accurately aligned with each other. To achieve this, first place a diaphragm with a hole about ⅛in (3mm) across accurately in its centre inside the eyepiece tube, and look through the eyepiece holder. If the instrument is accurately collimated you will see the image of the flat in the centre of the main mirror. (On larger instruments the image should be slightly displaced towards the top of the tube.) Then look at the flat mirror and you should see the main mirror with the flat at its centre, and in the centre of the system the image of your own eye.

Screws are usually provided for collimating a reflector, and after a little patience you should be able to true the system up satisfactorily (Fig 4).

To this, I add my own final check, as follows. Take an eyepiece of small magnification and select a bright star to be seen with it in the telescope. Focus the star accurately at the centre of the field. Then move the star say halfway towards the edge of the field to the right, when a coma may appear and the star begins to look like a comet with its tail. Now repeat the same setting to the left and check that the length of the 'comet' tail is the same length as on the right. Repeat this at both top and bottom of the field. If the telescope is properly adjusted, all comae or tails should be of equal length. If the eyepiece shows no coma, then the field is truly flat, and all you need to do is to check that the sharpness of the image is equal in all positions in the field.

It is perhaps unnecessary to remind readers again that both the main mirror and the flat of a reflector should be kept covered. The main mirror is safer if uncovered last before use and covered first after use.

Mirrors are nowadays usually aluminised, thus overcoming the bugbear of constant resilvering. It used to be quite usual for the amateur astronomer to resilver his mirrors himself, but today the aluminising is best done professionally. A number of optical firms do this work, and in the UK advice can be obtained by members of the British Astronomical Association from either the Curator of Instruments or the Director of the Instruments and Observing Methods Section. Nevertheless, it may be necessary to maintain the instrument in good order by occasionally cleaning the mirrors, which become tarnished or covered with a grey deposit. This deposit is noticeable when the mirror is viewed at a small angle, but is not so apparent when the mirror is seen face on, as it is when in use in the telescope. A mirror can look grey but at the same time work quite well, especially on the Moon and planets, where there is plenty of light coming into the instrument. With faint stars it is a different matter, and a mirror in first-class condition will show stars of far lower magnitude than one that is not.

It is remarkable how poor a mirror's condition may appear while yet the telescope works satisfactorily. The time does come, however, when something has to be done. Here care is required; one must not thin the mirror's film by over-cleaning, for then light may pass through it or through pores produced when cleaning it.

To wash a mirror, first take it out of its cell and stand it nearly upright. Using some cotton wool, stream water over the face to wash away any grit. After being satisfied that no grit at all remains, start to wash the mirror from the top edge, applying soap on cotton wool and working downwards across the face. When the mirror is clean, rinse the soap away and dry the mirror in the air naturally; or use a well washed handkerchief, soft to the touch, to dab the surplus water off.

The result may well be a mirror with streaks across it, but these can be gently polished off with a very clean, soft well washed handkerchief or cloth of similar soft material. An aluminised surface will withstand polishing better than a silver one, which is much more delicate.

The foregoing method is applicable to both the main mirror and the flat, provided they are aluminised. Silvered mirrors can be washed in a similar manner, but very great care is necessary to avoid damaging the silver film. Hot water should not be used.

After cleaning, the mirrors should be replaced in their cells in exactly the same position as before—to accomplish this, it is a good idea to mark the edge of the mirror and the cell before removing the mirror.

Note that mirrors should be a loose fit in their cells. Cardboard slips can be inserted to pack the mirror in its cell without any points pinching the glass, which may easily destroy its figure and so render the instrument useless for satisfactory observation. The fore and aft motion is usually checked either by a rim or by brackets holding the mirror in its cell.

These, again, should not be a tight fit, for obvious reasons.

If these precautions are taken and the mirror is a good one, the images of stars should be small and clean. If the mirror is pinched in any way, or is not perfect, stars may be shown outside the focus as oval, and within the focus as oval but with the longer diameter at right-angles to that of the outside image. This is either the effect of astigmatism, which may be inherent in the main mirror or the flat, or caused by pinching at some point.

It is well known that aluminised mirrors last much longer than silvered ones, especially if a protective coating is applied after aluminising. I have used an aluminised mirror for many years without any more attention than that described above, but it is always kept securely covered when not in use.

It is hardly necessary to mention that mirrors should not be touched with the fingers.

Time, the Clock and Position

Instructions on adjusting the equatorial mounting of a telescope have already been given. Now some explanation of the principles involved is necessary for full comprehension of what is behind these instructions.

Time is a measure of change. If everything stood still, including all movements, from the electrons to the galaxies, there would be no passage of time. The Universe would be truly static and eternal. But we live in a time-controlled world.

It is convenient to record time as the period of the revolution of the Earth around the Sun, so producing years; and the rotation of the Earth with reference to the so-called fixed stars (fixed because they are so far away that their individual motions can be ignored) and the apparent passage of the Sun over the meridian each noon, so producing days. There are thus two types of year and day, one referring to the stars and the other to the Sun. The first is called the sidereal period and the second the solar period, and similarly we have both sidereal days and solar days.

In normal daily life we use the solar day, but in astronomical work we must also use the sidereal day.

Each solar day is divided into hours, minutes and seconds on the civil clock, with midnight at Greenwich as the standard beginning of each day. This system is called Universal Time, and local times in New York or Melbourne, say, must be adjusted for astronomical purposes to this standard when recording each observation.

These days are referred to the passage of the Mean Sun over the meridian at noon. We cannot use the Sun itself because its period varies with the seasonal acceleration and slowing of the Earth in its orbit, so we average out the differences and call the resultant motion that of the Mean Sun. This is the 24-hour day used in our clocks for such civil affairs as train times.

During each solar day the Earth moves forward in its orbit, so that a day by the stars is of different duration from that by the Sun. The sidereal day (referred to the stars) is shorter than that referred to the Sun by about 3 min 56 sec. Thus the sidereal clock gains on the civil clock 3 min 56 sec each day. Another way of looking at it is to say that during the year the line joining the Earth and the Sun revolves once, while that referred to the stars stays constant in direction. The result is that there is one more sidereal revolution during the solar year, so that the sidereal day is shorter.

In the case of solar time we refer to the position of the Mean Sun, but in the case of sidereal time we have to refer to the fixed star background. The point of reference selected is not a certain star, but the point on the celestial sphere where the ecliptic (the circle traced on the celestial sphere by the apparent motion of the Sun during the year) crosses the celestial equator. This is called the First Point of Aries, because it was originally in the constellation of Aries; however, due to the precession of the Earth's orbit in space, this point moves and is now in the constellation of Pisces. This motion along the celestial equator amounts to 3·07 seconds of arc each year.

Each sidereal day begins when the First Point of Aries crosses the meridian of the place of the observer, and so is what is called local time. For observers west of Greenwich it is later, while for those east of Greenwich it is earlier. The sidereal day is divided into sidereal hours, minutes and seconds.

It is obvious that the sidereal time of the passage of a star across the meridian of the observer depends on the angle between the First Point of Aries and the position of the star. This brings us to the conception of what is termed Right Ascension. This is the arc on the celestial equator between the First Point of Aries and any particular star or position produced onto the celestial equator. This arc is not expressed in degrees etc., but in time, at the rate of one hour of time for each 15°. Thus Right Ascension is not only a determination of position, but is also the local sidereal time at which the star will cross the observer's meridian. It will be seen that since the Earth rotates once per 24 hr, 360° is equivalent to 24 hr, 180° equivalent to 12 hr, 15° equivalent to 1 hr, and so on.

To find a star's position in Right Ascension we take the local sidereal time and add or subtract so many hours, minutes and seconds according to the star's Right Ascension in the catalogue. This gives us the angle expressed in time by which the star is removed from the local meridian. This angle is known as the Hour Angle of the star, and is reckoned from the meridian towards the west. If the required star position is east of the local meridian it is sometimes convenient to set the telescope on the meridian and at the required angle above or below the celestial equator (that is, in declination, which we shall consider later on), note the local sidereal time and then wait for the time of the hour angle to elapse, when the star should be in the field of view.

For ease of reference, the relation between angles expressed in degrees and time intervals less than an hour is set out below:

Minutes of time	Degrees and minutes of arc	
45	11°	15'
30	7°	30'
20	5°	00'
15	3°	45'
10	2°	30'
5	1°	15'
4	1°	00'
3		45'
2		30'
1		15'

When using a telescope with setting circles, you can find an object by offsetting from the Right Ascension position of a known star. There is one little difficulty to be overcome when the circle of Right Ascension is marked with increasing values towards the west, as in Hour Angle. In this case the procedure is to set the instrument on a star whose Right Ascension is known, but to set the circle to read the Right Ascension of the object to be found. Then setting the telescope so that the circle reads the Right Ascension of the known star should bring the required object into view.

It is a convenience to have the circle marked with increasing value towards the east. Then all one has to do is to set the telescope on a star whose Right Ascension is known, set the circle to read that value, and swing the instrument until the circle reads the Right Ascension of the required object.

For this purpose it is useful to have a list of reference stars spread as evenly over the celestial sphere as possible, so that they can be used for finding other objects over a wide field. Such a list appears in Table 3, which includes the star's declinations, a term we should consider before using the list.

Declination is a measure of angular distance either north or south of the celestial equator. Thus stars on the equator have declination zero while the poles have declinations of 90° north and south respectively.

The full description of the position of a star is thus composed of two elements—Right Ascension, which we have already discussed, and declination north or south. These two terms are often designated by the Greek letters alpha (α) and delta (δ), or by the abbreviations RA and dec, and give a precise position for any celestial object.

Table 3

Short List of Reference Stars

Star designation	Right Ascension (1970)		Declination		
	hr	min		deg	min
Alpha Andromedae	00	6·8	N	28	55
Gamma Andromedae	2	2·0	N	42	11
Beta Persei	3	6·0	N	40	50
Alpha Tauri	4	34·2	N	16	27
Alpha Aurigae	5	14·5	N	45	58
Beta Tauri	5	24·4	N	28	35
Alpha Orionis	5	53·5	N	7	24
Gamma Geminorum	6	36·0	N	16	26
Alpha Geminorum	7	32·7	N	31	57
Alpha Canis Minoris	7	37·7	N	5	18
Alpha Leonis	10	6·8	N	12	07
Alpha Ursae Majoris	11	1·9	N	61	55
Beta Leonis	11	47·5	N	14	44
Eta Ursae Majoris	13	46·4	N	49	28
Alpha Boötis	14	14·3	N	19	20
Alpha Scorpii	16	27·6	S	26	22
Alpha Lyrae	18	35·9	N	38	45
Gamma Cygni	20	21·1	N	40	09
Alpha Cygni	20	40·4	N	45	10
Epsilon Pegasi	21	42·7	N	9	45
Beta Pegasi	23	2·3	N	27	55

The positions in Table 3 have all been updated to the year 1970, and should be near enough for ordinary finding purposes for many years to come. Over a period of 10 years the maximum change in Right Ascension is only 0·9 min while that in declination is 3′. For those requiring greater accuracy, Table 4 supplies corrections.

Table 4

Corrections to be Applied for Precession

Ten-year Precession in Right Ascension

RA	Declination						
	+60° (min)	+40° (min)	+20° (min)	0° (min)	−20° (min)	−40° (min)	−60° (min)
0ʰ	+0·5	+0·5	+0·5	+0·5	+0·5	+0·5	+0·5
1ʰ	+0·6	+0·6	+0·5	+0·5	+0·5	+0·5	+0·4
2ʰ	+0·7	+0·6	+0·6	+0·5	+0·5	+0·4	+0·3
3ʰ	+0·8	+0·6	+0·6	+0·5	+0·5	+0·4	+0·2
4ʰ	+0·8	+0·7	+0·6	+0·5	+0·4	+0·3	+0·2
5ʰ	+0·9	+0·7	+0·6	+0·5	+0·4	+0·3	+0·1
6ʰ	+0·9	+0·7	+0·6	+0·5	+0·4	+0·3	+0·1
7ʰ	+0·9	+0·7	+0·6	+0·5	+0·4	+0·3	+0·1
8ʰ	+0·8	+0·7	+0·6	+0·5	+0·4	+0·3	+0·2
9ʰ	+0·8	+0·6	+0·6	+0·5	+0·5	+0·4	+0·2
10ʰ	+0·7	+0·6	+0·6	+0·5	+0·5	+0·4	+0·3
11ʰ	+0·6	+0·6	+0·5	+0·5	+0·5	+0·5	+0·4
12ʰ	+0·5	+0·5	+0·5	+0·5	+0·5	+0·5	+0·5
13ʰ	+0·4	+0·5	+0·5	+0·5	+0·5	+0·6	+0·6
14ʰ	+0·3	+0·4	+0·5	+0·5	+0·6	+0·6	+0·7
15ʰ	+0·2	+0·4	+0·5	+0·5	+0·6	+0·6	+0·8
16ʰ	+0·2	+0·3	+0·4	+0·5	+0·6	+0·7	+0·8
17ʰ	+0·1	+0·3	+0·4	+0·5	+0·6	+0·7	+0·9
18ʰ	+0·1	+0·3	+0·4	+0·5	+0·6	+0·7	+0·9
19ʰ	+0·1	+0·3	+0·4	+0·5	+0·6	+0·7	+0·9
20ʰ	+0·2	+0·3	+0·4	+0·5	+0·6	+0·7	+0·8
21ʰ	+0·2	+0·4	+0·5	+0·5	+0·6	+0·6	+0·8
22ʰ	+0·3	+0·4	+0·5	+0·5	+0·6	+0·6	+0·7
23ʰ	+0·4	+0·5	+0·5	+0·5	+0·5	+0·6	+0·6
24ʰ	+0·5	+0·5	+0·5	+0·5	+0·5	+0·5	+0·5

Ten-year Precession in Declination

RA		RA	
0ʰ	+3′	12ʰ	−3′
1ʰ	+3′	13ʰ	−3′
2ʰ	+3′	14ʰ	−3′
3ʰ	+2′	15ʰ	−2′
4ʰ	+2′	16ʰ	−2′
5ʰ	+1′	17ʰ	−1′
6ʰ	0	18ʰ	0
7ʰ	−1′	19ʰ	+1′
8ʰ	−2′	20ʰ	+2′
9ʰ	−2′	21ʰ	+2′
10ʰ	−3′	22ʰ	+3′
11ʰ	−3′	23ʰ	+3′
12ʰ	−3′	24ʰ	+3′

Observing the Sun

At the outset it must be conceded that there is little original work on the Sun that the amateur can do which is not already fully covered by professional institutions. This does not mean that there is nothing that can be done to advantage, nor does it mean that observation of the Sun is a project devoid of interest; and there is also the offchance that an amateur, in the course of his regular observation, may discover something overlooked by the professionals.

There is a further point that must be reiterated, even if it seems like labouring the obvious: *the extreme danger of looking at the Sun directly.* Remember that the object glass of a telescope gathers not only light but also heat. This is focused by the lenses and at the focus the intensity of both light and heat is greatly enhanced. Many a careless solar observer has set fire to his tie, and even a beard or two. I regularly light my pipe in the focus of a 4in (100mm) refractor when observing the Sun. This light and heat is, if the observer is careless or foolish enough, received by the eye with disastrous results. Even if the so-called 'solar filter' is used, there is still considerable danger, for filters have a habit of cracking or even exploding under the stress, and in accordance with 'Spode's Law' this will inevitably happen at the worst possible moment! (A filter on the objective is safer than one in the eyepiece.)

The *only* safe way to look at the solar disk is to project it onto a white screen, and then to draw or photograph the image.

It is also dangerous to use a reflecting telescope of any large size when observing the Sun because of the danger of cracking the flat mirror. Therefore the general procedure is to use a refractor of small or medium dimensions. A 4in (100mm) diameter objective is quite adequate for this purpose, and smaller sizes are not to be despised, since even a 2in (50mm) diameter glass will show the larger sunspots quite well by projection.

While dealing with this aspect of the subject, we should mention the Herschel wedge eyepiece for solar work. In this device the sunlight is split, and only that part of it reflected from the face of a narrow-angle prism is sent into the eyepiece, the main portion of the light, and heat, being passed through the prism and deflected away from the focus. This apparatus, sometimes called the solar diagonal, reduces the light and heat entering the eyepiece to about one-sixteenth of its original strength, but even then a further reduction is necessary for eye tolerance, and a filter is sometimes used. A green filter of sufficient density is best for ease of viewing detail, but great care is necessary to avoid damaging the eyes. We repeat—*the only safe way to observe the Sun is by projection onto a screen.* Damage to the observer's eye can occur if the wrong filter is used, even with the solar diagonal. An infra-red filter, for instance, passes the harmful long-wave radiation. For the safer way of projection use a long focus non-cemented eyepiece.

By the projection method the size of the image on the screen can be adjusted by moving the screen away from the eyepiece, thus enlarging the diameter. A smaller image may be obtained by bringing the screen to a point nearer the eyepiece. The best image for solar work is that with a diameter of about 6in (150mm). Smaller sizes render the spots a little difficult to locate, and large sizes tend to be unwieldy. The usual manner of mounting a screen behind the eyepiece is to use an adjustable rod (Fig 5).

On examining the solar disk on the screen, one will immediately notice the limb darkening (i.e., the darkening around the edges of the solar disk). In this darker area bright parts will often be seen; these are the faculae, which do not show up on the brighter part of the disk (Plate 9).

Often dark areas caused by the presence of sunspots will be observed, and here is a field wide open to the amateur observer. The first problem is how to draw or photograph these spots accurately. While photographs of the screen may be taken, the camera must be located as near as possible to the centre of the light beam from the telescope, as otherwise the Sun will appear oval in the picture. The more usual method is to draw what one sees, and an established method has been worked out for doing this.

Firstly, a good bright image of the Sun must be assured, and it is often found profitable to make a balsa wood box, painted black inside and with one side open, through which the screen can be viewed. This is

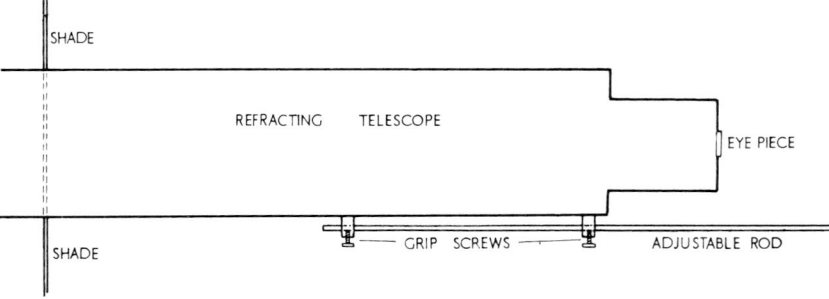

fixed to the eyepiece end of the telescope. There may be some difficulty here in getting a true impression of the image, as the observer is necessarily viewing from the side. If a screen is used, with a large masking sheet of some light material through which the telescope projects, this will shade the screen sufficiently for a good image to be seen, and the observer will be able to view it from nearer the centre of the emergent beam. It is a matter of taste or personal preference. This method is illustrated in Fig 5.

The screen on which the image of the Sun is projected should be ruled into squares, which should then be subdivided by diagonals (Fig 6). This gives one a very accurate system of location. Then draw a duplicate but darker diagram than the projection screen diagram, and place it under a semitransparent sheet of paper, on which the drawing is to be made. Draw a circle of 6in (150mm) diameter on the drawing paper, and arrange the projected solar image so that it fits the diagram on the screen.

The next step is to let the Sun drift across the screen by the Earth's rotational motion. This will fix the east-west line; the drift, of course, is in a westward direction. We now have to establish the north and south points at 90° from the east and west points. It is best to draw the two diameters involved on the drawing paper in readiness for observation, and the projection screen diagram can be rotated until the Sun appears to travel accurately along the east-west diameter of the disk. With an altazimuth mounting it is necessary to adjust the orientation frequently.

When all this has been accomplished, begin to draw in the apparent positions of any sunspots you may see, using the diagram under the semitransparent drawing paper to enable you to place the spots accurately by reference to the similar diagram on the screen. Having established the positions, usually by drawing in the dark central umbra of each spot, start shading in the details of the penumbra. The central umbra usually appears purple-brown in colour, and the penumbra a lighter shade of brown. This, of course, is a contrast effect, for the spots themselves are

Fig 5 Solar projection screen on a refractor telescope

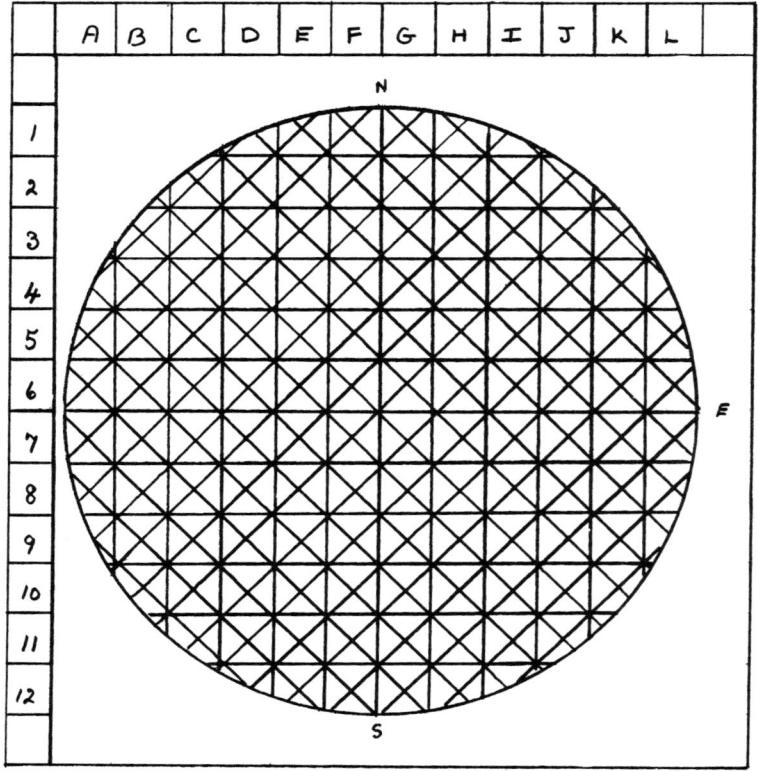

Fig 6 Solar disk graticule (*Mrs. J. Merrilees*)

really highly luminous. When the drawing is complete, you will have a record of the appearance of the Sun at that particular day and time. Record these details and those of the telescope used.

Now we come to the more difficult problem of establishing the relative positions of the sunspots in terms of their solar latitude and longitude, which are not constant in their relation to the position of the Earth. The orbit of the Earth is inclined to the solar equator, with the result that sometimes we see the Sun slightly from the north and sometimes slightly from the south. Sunspots therefore trace out very different paths at different times of the year (Fig 7).

The problem is now that of transforming the observed positions that have been recorded, relative to the Earth's position in its orbit, to the true positions on the solar globe.

The three following quantities must be established for the day of observation to enable this to be done:

40

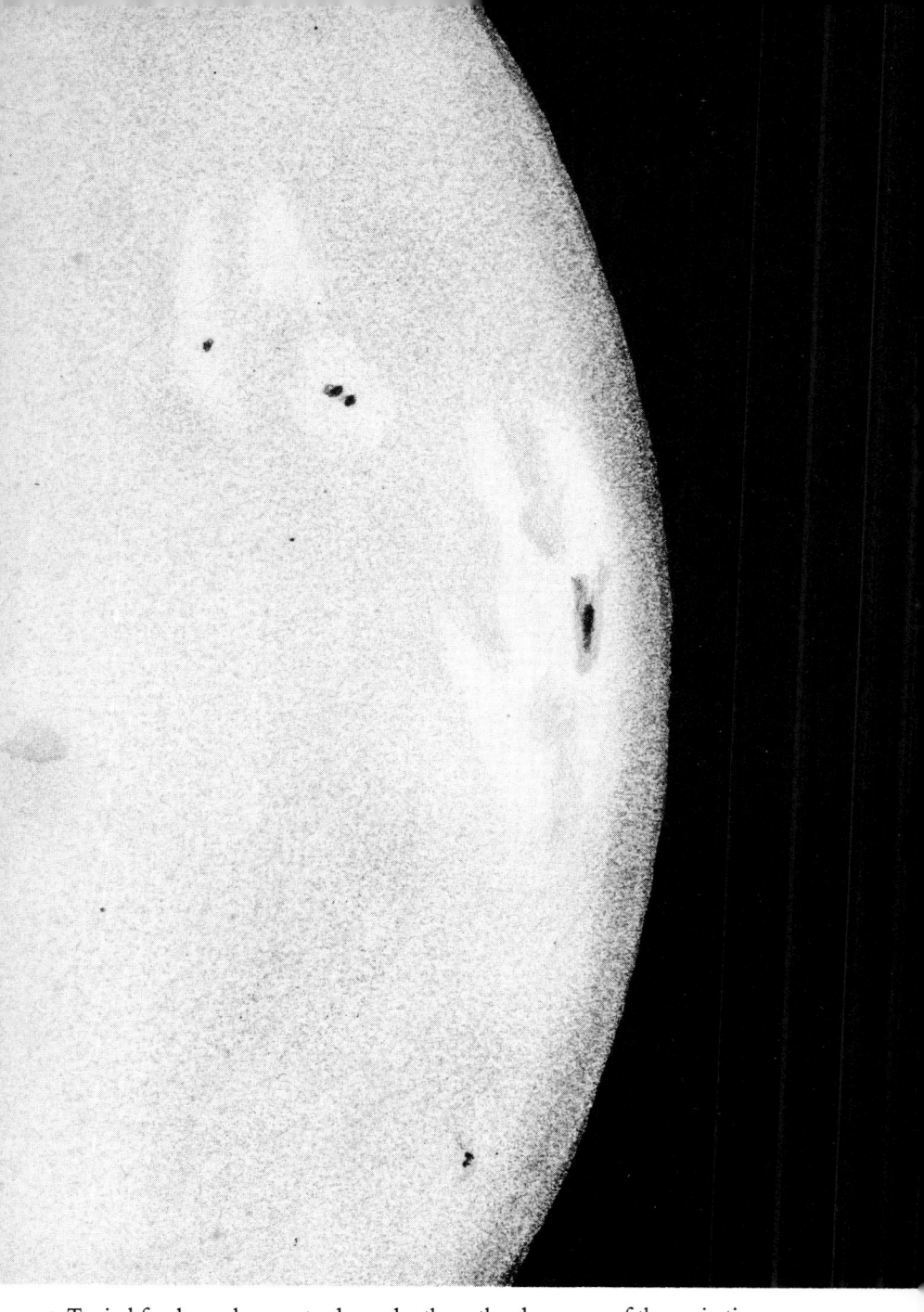

9 Typical faculae and sunspots, drawn by the author by means of the projection
method

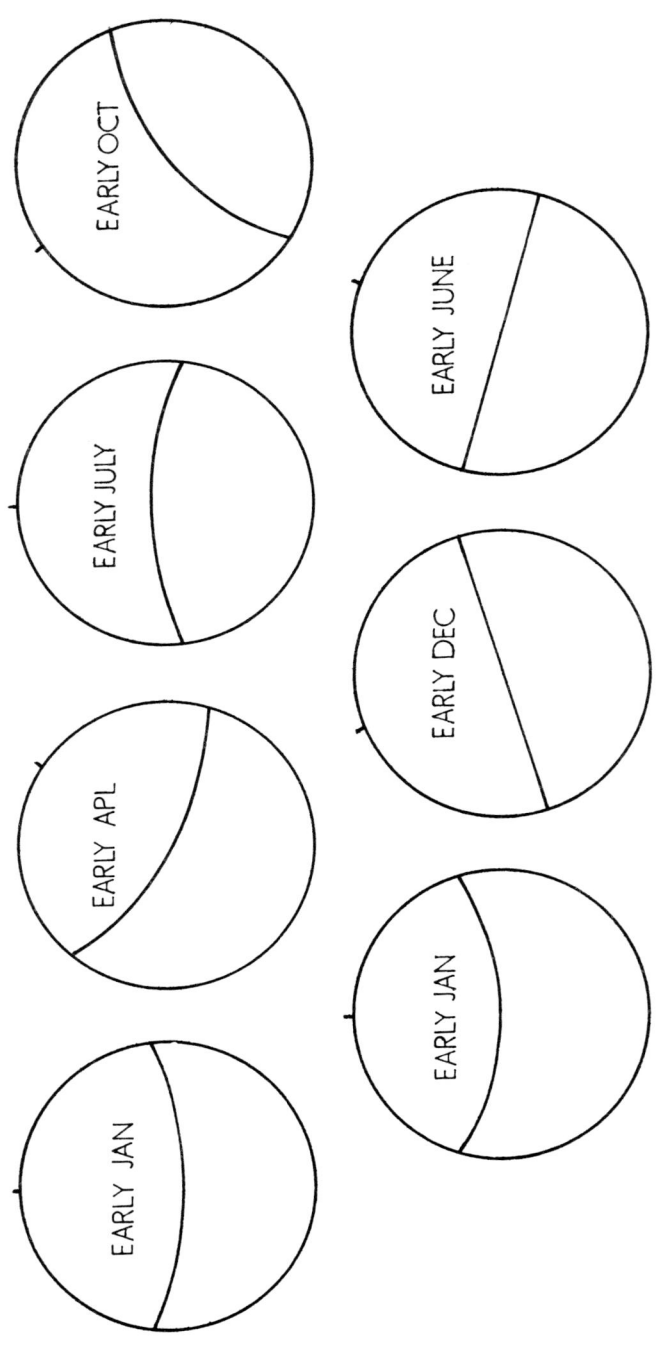

Fig 7 The apparent paths of sunspots across the solar disk change because of the apparent shift of the Sun's axis as the Earth carries the observer round the Sun annually. The apparent paths are straight lines only during early June and December

P The position angle of the north end of the solar axis of rotation. This is regarded as positive if east of the north point of the disk, and negative if west. Position angle is measured from the north through east, in degrees.

B_0 The heliographic latitude of the centre of the solar disk.

L_0 The heliographic longitude of the centre of the solar disk.

These three quantities are to be found in the *Astronomical Ephemeris* and the *Handbook* of the British Astronomical Association.

Since the Sun is a globe, it is clear that it is not just a matter of moving the drawing over a ruled disk whose rules represent on a flat plane the lines of solar latitude and longitude. The mathematics are a little difficult for those not at home with them. (In Britain, to ease matters, a diagram has been published, with full detailed instructions for its use, to enable one to read off the values needed by placing the drawing on its semi-transparent paper over the scales, and calculating the required positions. 'Porter's Disk' is obtainable from the British Astronomical Association, Burlington House, Piccadilly, London W1V 0NL.)

A long series of observations of this nature will provide material usable in detecting the drift of sunspots both to and from the solar equator, and also in longitude. Plotting such observations over a number of years, in latitude against date, will provide a confirmation of the well known Butterfly Diagram showing, with an 11-year cycle, the way that sunspots appear near the beginning of the cycle at higher latitudes, later on appearing progressively nearer to the solar equator.

Another line of observation is that of spot counts. In this programme the number of active areas of spots whose maximum distance apart is 10° is counted, together with the number of spots in each area. Then the number of spots plus ten times the number of active areas will equal the 'activity level'. To this must be applied a constant to represent the ability of the telescope, atmospheric conditions, and the acuity of the observer to see spots.

To find this personal constant one carries out a number of such counts, and then compares the figures published from Zurich for the days on which observations were made. The two sets of figures are adjusted by the required amount to make them agree, this adjusted figure becoming the observer's own personal constant. Thereafter the personal constant can be applied to observed counts directly, and in theory the results should agree with the Zurich figures, which are published in the US *Sky and Telescope* each month. The constant should be found to be close to unity, but will vary from observer to observer. It is usually applied as a factor by which the observed counts have to be multiplied to bring them in line with Zurich.

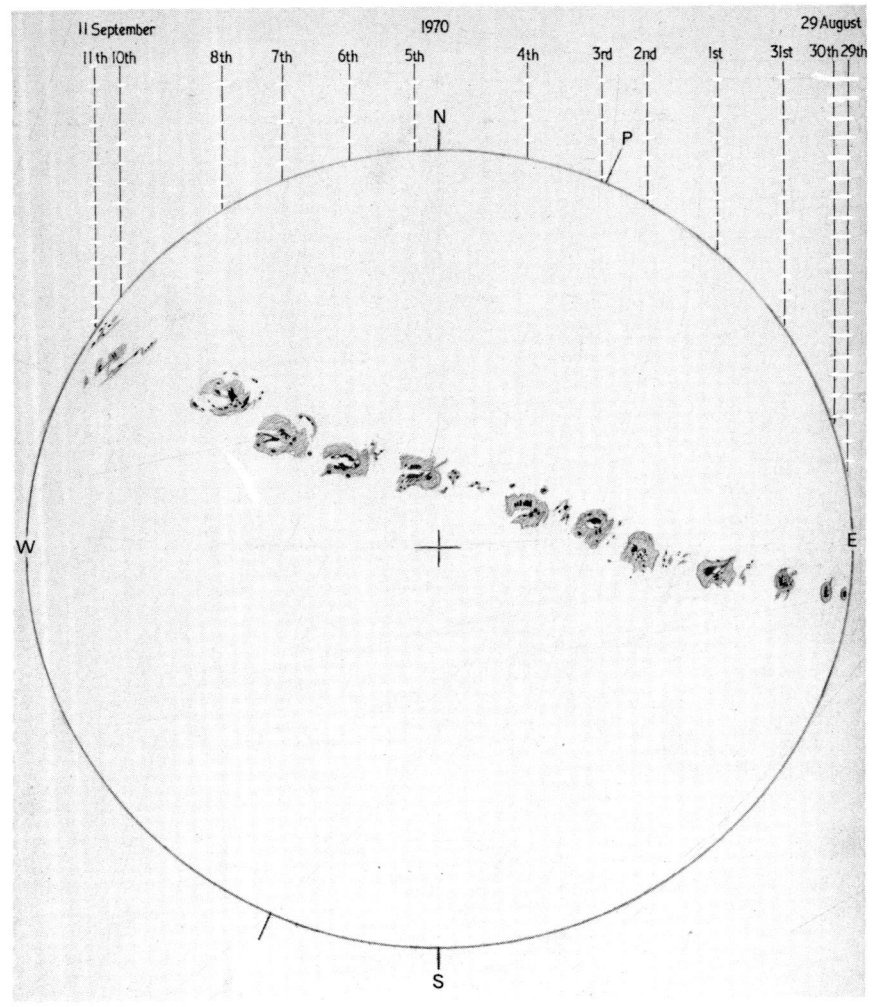

10 Composite drawing showing apparent movement of spots through solar rotation (*Mrs. J. Merrilees*)

11 *Right*, solar eclipse of 1973 June 30, showing corona (*C. J. R. Lord*)

Repeated observation of sunspots will demonstrate their rapid changes in appearance. The Wilson Effect, caused by foreshortening near the limb, will normally be apparent (Plate 10).

There is room for study of the behaviour of sunspots over their life-time, especially if records are accurately kept, so that a spot may be recognised on its return after passing around the far side of the Sun.

Detailed drawings can be of value in tracing its life history.

An observer who has a suitable spectroscope can make a study of the prominences, especially in the Hydrogen Alpha line. The spectroscope should be powerful enough to split the D lines of sodium for this work; and with, say, a 4in (100mm) objective or larger, if the slit is kept accurately at a tangent to the limb (and this is important) and slightly opened, then a prominence, if present, may be seen. Concurrence with radio fade-outs or aurorae makes a fascinating study.

Solar photography is dealt with under astronomical photography (page 23) except for photography of total solar eclipses, which offers opportunities to study the structure of the corona. From the central line of the eclipse on Earth satisfactory pictures can be taken with varying shutter speeds to obtain records of the inner and outer corona (Plate11). Such pictures give more accurate records than sketches, which, on account of the short duration of the event, have to be drawn too quickly to be considered reliable.

General Introduction to Lunar and Planetary Observation

The refractor telescope tends to give steadier images than the reflector, but the greater diameter of the average reflector gives better resolution when the air is steady.

Seeing conditions are classed on the Antoniadi Scale as follows:

I   Perfect seeing without a quiver;
II   Slight undulations, with moments of calm lasting several seconds;
III Moderate seeing, with large air tremors;
IV   Poor seeing, with constant troublesome undulations;
V   Very bad seeing, scarcely allowing the making of a rough sketch.

Roman figures are used to avoid confusion with other data.

Magnification used is recorded as, for example, X100 or X230, which mean, respectively, magnifications of 100 or 230 diameters.

Instruments are described by, for example, 150mm O.G. (which means a 150mm diameter object glass), or 200mm spec (which means a 200mm diameter reflector). Spec comes from the Latin *speculum*, meaning mirror.

Focussing is best done on the limb of the planet or Moon, which should show a sharp edge.

Observing the Moon

The advent of spaceflight, culminating in the landings on the Moon, has generated a feeling that Earthbound observation of our satellite is now useless. This is not really true. What is true, however, is that the type of observations carried out with Earthbound telescopes has changed dramatically during the past few years.

Time was when drawing details of lunar formations as seen at the telescope was of prime importance. Today it remains the best way to learn one's way about the Moon, and at the same time to get to know many of the details intimately, so that one can spot any apparent change or unusual feature immediately.

Transient Lunar Phenomena
Surveys of the surface have been completed for the major part of the lunar surface, but there remains at least one field of study wide open to the amateur observer or the professional who may be so inclined. This is the observation of Transient Lunar Phenomena, or TLPs.

In this connection it is recommended that the observer first gets to know some of the formations intimately by close inspection, backed up by drawing under many angles of illumination during the lunation. Once this has been achieved, one can start patrolling for TLPs. As the name implies, these are temporary phenomena, and they tend to occur in or near craters on the margins of the maria. Aristarchus, Gassendi, Plato are examples (see Figs 8–11). They are known to take the form of

Fig 8 The Moon: south-west quarter (*Mrs P. Helm and Dr P. Moore*)

Fig 9 The Moon: south–east quarter (*Mrs P. Helm and Dr P. Moore*)

Fig 10 The Moon: north-west quarter (*Mrs P. Helm and Dr P. Moore*)

Fig 11 The Moon: north-east quarter (*Mrs P. Helm and Dr P. Moore*)

obscurations or glows, and many are bright in red or blue light.

Search is carried out by patrolling with colour filters, usually mounted between the eyepiece and the eye in a holder that permits rapid change from, say, red to blue and back again. This device, which is called the Moonblink device (Fig 12), should be used with a magnification of about 200 diameters on a telescope of some 8in (200mm) diameter as a minimum.

A common alternative is to mount the filters in a disk that can be rotated rapidly by suitable gearing, so as to expose the filters in succession as required. In this form the filters are usually mounted in front of the eyepiece so as to get the lens as close to the eye as possible, with the filter beyond it, out of harm's way enclosed in some form of casing into which the eyepiece fits. The usual filters employed are the Kodak Wratten series numbers 25 (red) and 44A (blue). These are matched as nearly as possible for visual light transmission, so that the eye can adjust rapidly to the change of wavelength. They are obtainable in gelatine much more cheaply than in glass.

The technique is to look carefully at a particular formation in, say, red light and then switch to blue, and by memory compare the images, so that any bright area in red light that does not appear in blue will be quickly spotted. Conversely, any area bright in blue and lacking in red response will also be noted.

Sometimes a TLP may be colourless, and then it is that detailed knowledge of the formation comes into play. Any unusual brightening or any obscuration should then be noted.

These phenomena normally last a few minutes with an upper limit of about three-quarters of an hour or so, although sometimes a TLP may last for a longer period; if it is long-lasting, however, one should become suspicious, since the appearance may easily be a normal effect that one has overlooked previously. Examples of this are the lack of detail that is sometimes evident on the mountain range within the crater Gassendi, and the often seen ruddy hue of the northern part of the floor of Fracastorius. The floor patches in Plato are usually more evident in red light than in blue; these may be regarded as permanent 'blinks' (so-called because they blink when the filters are switched from red to blue and back again).

When patrolling the lunar surface in this way, keep a special watch on Alphonsus, for here in 1955 Alter noted a blurred effect in blue and violet light, and Kozyrev in 1958 photographed a reddening of the central peak, which faded to its normal appearance shortly afterwards. This was the first occasion on which a photograph of a TLP was obtained, and has become the classical case verifying the reality of such phenomena. Since then many TLPs have been recorded in a number of craters, both inside and out.

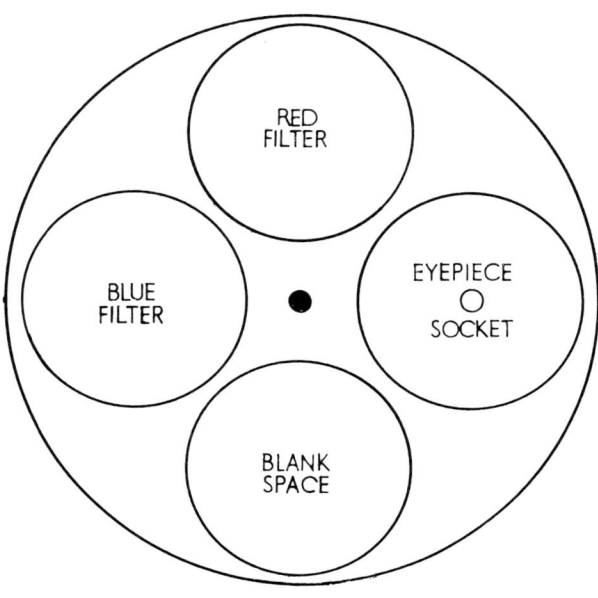

FACE-ON VIEW

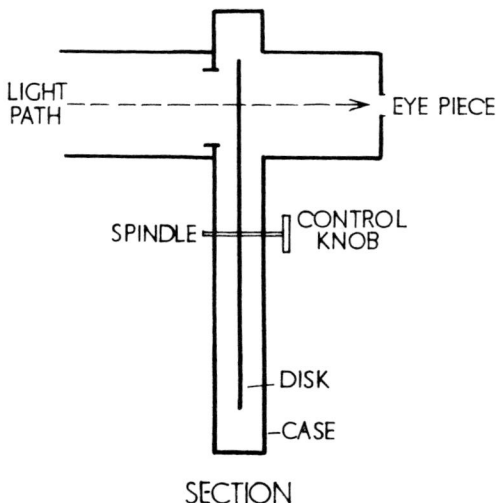

SECTION

Fig 12 Moonblink device

53

A word of caution is necessary here; I had the opportunity of observing the area in Alphonsus shortly after Kozyrev's observation with a 24in (610mm) reflector, when extremely transient moving patches of colour were to be seen. This was, of course, an effect of the Earth's atmosphere. Spurious TLPs are quite common, and it is usual, when a TLP is suspected, to contact an observer at some other station for confirmation. If the event is found to be observed by both, then the network of observers is alerted by telephone in the hope of obtaining full coverage of the event. This network includes observatories equipped with spectroscopic apparatus, and it should prove possible to obtain spectrograms of the TLP that may explain it.

When reporting such events you should state whether the appearance was bright or dark in red, blue and integrated white light, and whether full detail was seen or lost, and give the date and time of the event. These details are required for analysis, which may uncover the cause of the TLP.

A typical record of a TLP follows:

1976 October 4

20 hr 55 min (about) Gassendi outside W (classical) wall blinks bright in red, normal in blue.

20 hr 58 min (about) Gassendi normal again.

21 hr 10 min (about) Another observer alerted me by phone asking for confirmation of blink in Gassendi.

A third observer alerted by me.

21 hr 21 min Gassendi normal.

21 hr 50 min Third observer telephones reporting possible obscuration over S part of Gassendi.

21 hr 52 min Fourth observer telephones confirmation of blink.

22 hr 00 min Gassendi appears normal again.

Altitude of Moon above horizon 33 deg.

The times quoted as '(about)' are so shown because accurate timing was not accomplished until 21 hr 21 min through clock error. The description '(classical)' refers to the old reckoning, with Mare Crisium considered as near the western limb. This has recently been reversed to make Mare Crisium near the eastern limb. When reporting it is necessary to be quite clear which system is used. The modern one agreed by the IAU is in accordance with the Apollo flights and was adopted by NASA.

Occultations

A second project in the modern programme of lunar observation is the recording of occultations—the disappearance and subsequent reappearance of stars when the Moon passes in front of them. These events may take place at either the light or the dark limb of the Moon, according to its phase, or age. Combined observations from two or more stations can yield both an accurate position of the Moon, and at the same time, a measure of the rotation period of the Earth.

Occultations are predicted for various positions on Earth and are published annually. These predictions are made for standard stations, and the times must be adjusted in accordance with the latitude and longitude of the observer.

The formulae for these corrections are applied to the original standard times in the following form: the time at the observer's station equals the standard station's time plus or minus the difference in longitude, plus or minus the difference in latitude. These functions of adjustment are considered plus if the observer is north of the standard station, and minus if he is south of it; and plus if he is west of the standard station and minus if he is east.

The coefficients of the adjustments for latitude and longitude are published in the ephemeris in the *Handbook* of the British Astronomical Association and elsewhere, and the standard stations are given as in Table 5.

Table 5

Place	Latitude (°)		Longitude (°)	
Greenwich	N	51·5		0·0
Sydney	S	33·9	E	151·2
Dunedin	S	45·9	E	170·5
Edinburgh	N	56·0	W	3·2
Melbourne	S	37·8	E	145·0
Wellington	S	41·3	E	174·8

Observations of occultations can be carried out with telescopes of quite moderate sizes, a 3in (75mm) refractor being quite adequate. There is often trouble with the glare of the Moon drowning faint stars, though the degree in which this trouble interferes with observation is somewhat dependent on atmospheric clarity. A little mist can make things difficult.

The position of the observer must be accurately known to 1 second of arc; that is, about 100ft (30m) on the surface of Earth. Timing, which must also be accurate, can be carried out by the use of a stop watch combined with radio time signals, to an accuracy of a tenth of a second.

Radio time signals are regularly broadcast by the stations listed in Table 6.

Table 6

Station	Frequency (kHz)
Rugby (England)	60, 2,500, 5,000, 10,000
Lyndhurst (Australia)	4,500, 7,500, 12,000
Fort Collins (USA)	2,500, 5,000, 10,000, 15,000, 20,000, 25,000
Ottawa (Canada)	3,330, 7,335, 14,670

One can also use the telephone speaking clock where available.

When an occultation occurs, the star usually disappears or reappears instantaneously, since it is a point of source of light. Occasionally, though, the star seems to fade and then be extinguished, or reappear and then brighten. These fading occultations are thought to be caused by the star being a binary, one component of which is occulted after the other (reappearing in like manner). The degree of fade obviously depends on the angle made by the line joining the stars with the limb of the Moon. The maximum duration of fade occurs when the stars' position angle is such that it allows the longest time lapse between the two occultations. At the leading and following limbs of the Moon this will be a right-angle to the lunar limb. At other points on the lunar limb the situation is different; since the Moon moves eastwards in its orbit there is a tangential effect away from the east-west line that becomes greater as one moves towards the north or south limb. A number of stars have recently been found to be binary systems through observation of occultations in this way.

We should also consider the case of grazing occultations, where the occulted star is placed so that it is only just occulted, near the north or south limbs of the Moon. On these occasions the light of the star can be seen to wink on and off as it passes behind the lunar mountains only to reappear in the following valley. It will usually be necessary to travel to a suitable point on the Earth to make such observations, but the recorded times (to the nearest second, if possible) are very valuable in the study of the shape and motion of the Moon.

Eclipses of the Moon are of interest to the modern observer because of the irregular brightening of certain lunar areas when immersed in the Earth's shadow. This is akin to TLP-observing, and may be carried out in the same way.

Observing Mercury

Mercury is an interior planet; that is to say, its orbit lies wholly within that of the Earth. Because of its proximity to the Sun, its maximum angle of elongation is only some 27° 45' (see figure 13). Mercury occasionally passes between the Sun and Earth and is seen to transit the disk of the Sun as a small black spot, with a maximum angular diameter of 10·5".

It will easily be appreciated that Mercury can only be seen well by the naked eye just before sunrise or just after sunset. In these conditions, seeing low down near the horizon is usually too poor for accurate observation, but one has to make the best of the situation.

It must be stressed that finding Mercury near the Sun with a telescope not provided with position circles is an undertaking that should be attempted with great care. It is all too easy to swing the telescope too near the Sun, and so receive into the instrument the blinding light that can do serious damage to the observer's eyes. When sweeping in this manner, it is advisable to start as near the Sun as conditions will permit, and to swing the instrument away from the Sun, so avoiding the danger.

A very satisafactory way is to note the position of Mercury relative to terrestrial landmarks one day, and use this position as a starting point for sweeping on the following day in the sunset daylight, allowing for the time difference at 15° per hour. If evening observation is intended, try to catch Mercury as soon as it appears in the twilight sky. Morning observation enables one to see Mercury in a relatively dark sky before sunrise, and then to follow it into the brightening sky as the Sun rises.

Sweeping is best done with a low-power eyepiece, which will give a large field of view. The telescope can be approximately focused in advance, say on a landmark on the horizon. Start the sweep some little distance above or below the known approximate position, move the telescope slowly over the area, then depress or elevate the instrument about half the diameter of the field of view, and make another pass over the same angle. Repeat this process until the planet is located in the low-power eyepiece. Then centre it and transfer to a higher power for a closer examination.

With circles, it is not necessary to use a clock for local sidereal time, but it is quite possible to set the telescope by offsetting from the Sun in the manner described on page 34. The Right Ascension of the Sun can be set up by observing the shadow of the telescope, or the bright exit pupil of the finder, if one is fitted. In any case, whatever you do, *do not look through the main telescope or even the finder to locate the Sun*. If you desire to find the position of Mercury before sunrise, follow the same procedure to offset from a bright star. It will be noted that the mention of offsetting from the Sun presumes the observation is to be made in daylight.

With adequate telescopic power, say with an instrument of some 12in (300mm) diameter, and when Mercury is suitably positioned relative to the Sun, it is possible to locate and observe the planet in full daylight. Under these conditions the planet is higher above the horizon and seeing should be better than when it is low down. There is a disadvantage in

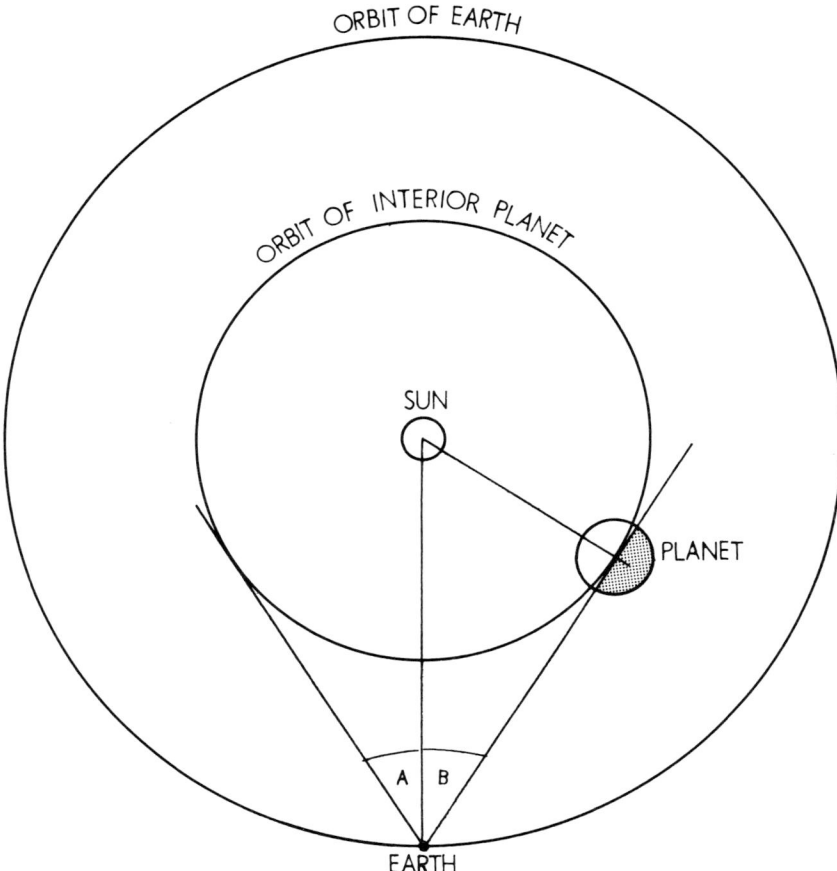

Fig 13 Elongations of an interior planet. A and B are the angles of elongation. Dichotomy occurs when the angle Sun-Planet-Earth is a right-angle

that Mercury is always near the Sun and, when it is high up, the Sun is also high in the sky. This entails poor seeing through disturbance of the air, especially near noon, by the solar heat. It would be better to try in the early morning when the air is cooler and less disturbed.

Once found, it is often surprising to note how small the apparent disk of Mercury appears in the telescope. When it is nearest to the Earth, the planet subtends a diameter of only 10·5″, and in other parts of its orbit this figure is proportionately less. At elongation the apparent diameter is only 7–8″. In comparison, the Moon subtends an angle of 30′, some 260 times as much. It is thus quite easy to overlook Mercury when sweeping for it with a low power.

Near elongation the planet appears almost exactly cut in two, with the terminator between the light and dark sides being almost a straight line. This cutting in two is called dichotomy (Fig 13). As the orbit of Mercury is elliptical, dichotomy is a little irregular in comparison with its elongation, so that we say it appears dichotomised when near elongation and not necessarily *at* elongation.

A telescope of medium size will show the phases, but with larger instruments of, say, 10in (250mm) diameter the phase estimates will be more accurate, and hence of greater value.

The usual method of recording the phases is by drawing to a scale of 2in (50mm) to the diameter of the planet. Draw a circle 2in (50mm) diameter on a piece of paper, or if you are a methodical observer (and all observers should be methodical), in the notebook. Examine the planet for a few minutes to allow the eye to adjust to the conditions, and draw a first impression of the line of the terminator on the prepared circle. Turn the paper through 90° and your head to match, and adjust the terminator on the drawing accordingly. This course is adopted because the eye is more accurate at deciding shapes than merely the position of a line. One cannot be too careful in drawing the terminator of a planet, for its position affects the whole appearance and, if care is not exercised, when one comes to insert any disk details the whole picture can easily look wrong. So it is necessary to get the terminator right in the first place. This records the phase, which is reckoned as a percentage of the diameter of the full disk of the planet.

After recording the terminator, look closely at the bright portion of the planet and note if there are any dark shadings visible; if so, draw them in. To see these reasonably well, you will need a telescope of at least 10in (250mm) in the case of a reflector or 6–8in (150–200mm) for a refractor, though the larger the aperture the better. Small telescopes are useless for recording dark areas on Mercury, and working with them only leads to confusion when it comes to analysing the drawings (Plate 12).

We know from space-probe observations and photographs that the

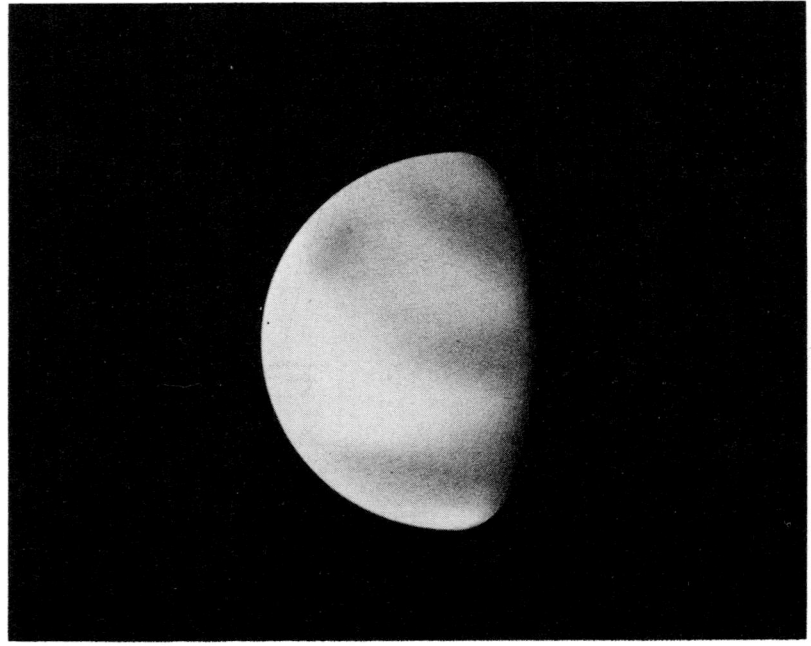

12 Drawing of Mercury 1977 April 22, 16 hr 30 min to 17 hr, with 16·5in (419mm) spec, X248 and X372 (*P. B. Doherty*)

surface of Mercury has craters on it similar to those on the Moon, but these cannot be seen from Earth. We do know that dark areas visible from Earth have been recorded by space probes. The light and dark areas are mapped on what is known as an albedo map, albedo being a measure of the reflectivity of planetary surfaces or surface features. These dark areas appear to repeat their appearances at elongations east and west of the Sun regularly. This regularity arises from the fact that Mercury's rotation on its axis once every 59 days happens to tie in with its synodic revolution so that we see one part at eastern and another at western elongations.

If, as was generally supposed until recently, Mercury has no atmosphere then these dark areas would always appear of the same shape and in the same positions. In addition, one would expect dichotomy to occur at precisely calculated times.

Mariner 10 has shown the existence of a tenuous atmosphere, unable to hold matter in suspension, on Mercury, but changes in its appearance from time to time must be watched for; and there is also room for a lengthy study of the times of dichotomy in relation to the orbital position of the planet.

In watching for these irregularities, you will find either a yellow or orange filter a help, for both tend to clarify the image. The apodising screen (see page 28) has also been found a useful accessory.

When Mercury appears to transit the disk of the Sun (the next time will be in 1986 on 13 November at about 4am Universal Time), it sometimes shows a halo, dark when near the centre of the solar disk, and light when near the solar limb—as it did in 1973, as shown by the beautiful drawings by R. M. Baum (Plate 13). This may well be evidence of an atmosphere surrounding Mercury. Transits of Mercury are most safely observed by use of the projection method described for solar observations. There is plenty of work waiting to be done on this little planet by those with adequate instrumental apparatus.

Observing Venus

Venus, like Mercury, is an interior planet. Like Mercury, too, Venus shows phases and differences in apparent size, according to its place in orbit relative to that of Earth. Here the similarity ends, for Venus is much larger than Mercury, having a diameter of 7,600 miles (12,10ckm); and, on coming much nearer Earth at inferior conjunction, Venus is a magnificent sight in even a small instrument at this time.

The larger orbit of Venus than that of Mercury carries Venus further away from Earth at superior conjunction, when it is on the far side of the Sun from Earth. This means that the apparent diameter of Venus as seen in the telescope varies considerably. At inferior conjunction, when Venus lies between Earth and Sun, the apparent diameter can reach a maximum of 63″, but when at its furthest from us Venus presents a disk only 10″ in diameter, which is still equivalent to the largest diameter that Mercury can present. It is apparent, therefore, that Venus is a much more rewarding sight in the telescope than ever Mercury can be. With Venus, of course, we do not see the surface, but only clouds and their shadows.

When near inferior conjunction, Venus shows a beautiful crescent of

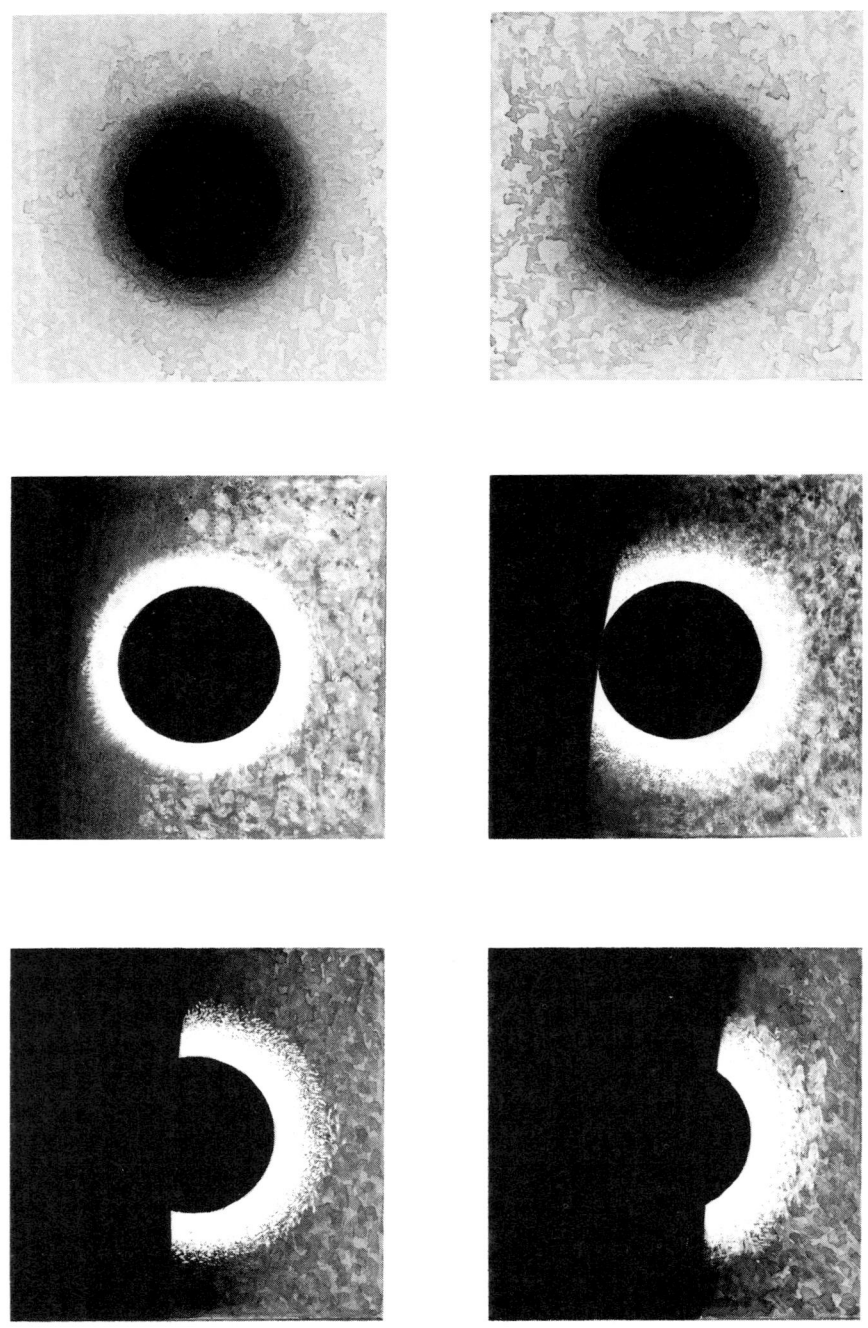

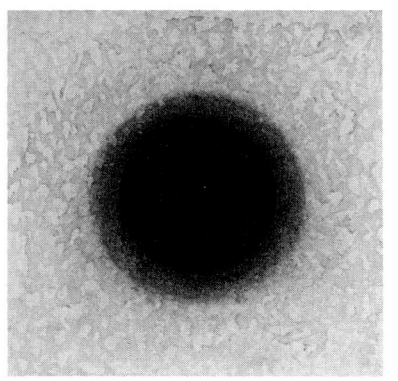
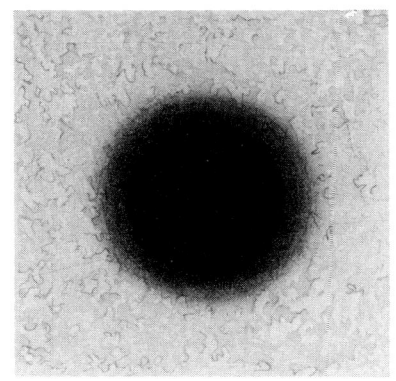
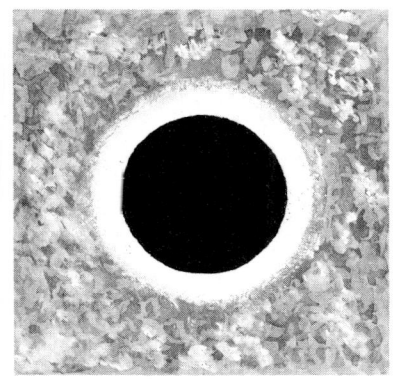
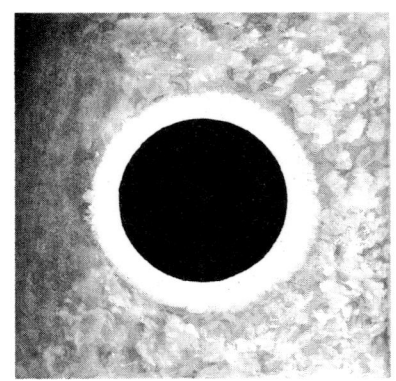

13 Transit of Mercury 1973
November 10, showing the halo dark
against the Sun's photosphere and bright
when nearing the solar limb
(*R. M. Baum*)

maximum size, and on reaching dichotomy on either side of the Sun (east or west) still measures some 25″ in diameter. The gibbous phase is necessarily smaller than this, as the planet is then further removed from Earth.

The brilliance of the planet depends not only on its distance from Earth but also on the phase illuminated. Maximum brilliance occurs between 34 and 37 days either side of inferior conjunction. This great brilliance makes the planet easily visible in daylight, and this is the best time to observe it. After dark in the evening, or before it gets light in the morning, the brilliant glare in the telescope makes serious observation quite impossible, unless some means are taken to reduce the light received. This is not really satisfactory, and so it is by far the better plan to observe Venus in daylight.

Venus can be picked up by the telescope in much the same way as Mercury, using the Sun as a marker for Right Ascension and setting the instrument in declination. On the other hand, it is usually easy to find Venus in the morning and evening twilight, and to start observation before it gets too low in the evening sky. In the morning there is plenty of time to spare; and as the day gets brighter conditions will improve, until one will find oneself observing in broad daylight.

Perhaps the most observed phenomena are the changes in the phase of Venus. Here, mere estimates are notorious for error. The best way, unless one can use a micrometer, is, as with Mercury, to draw the phase of the planet on a circle of 2in (50mm) diameter, then rotate both the drawing and your own head through 90° so as to view the planet at another angle. Thus you can compensate for estimating errors and finally come to an accurate decision as to the exact position of the terminator. The phase is then represented as the percentage of the whole disk illuminated, along a line at right angles to the line joining the points of the crescent or apparent poles of the gibbous phase.

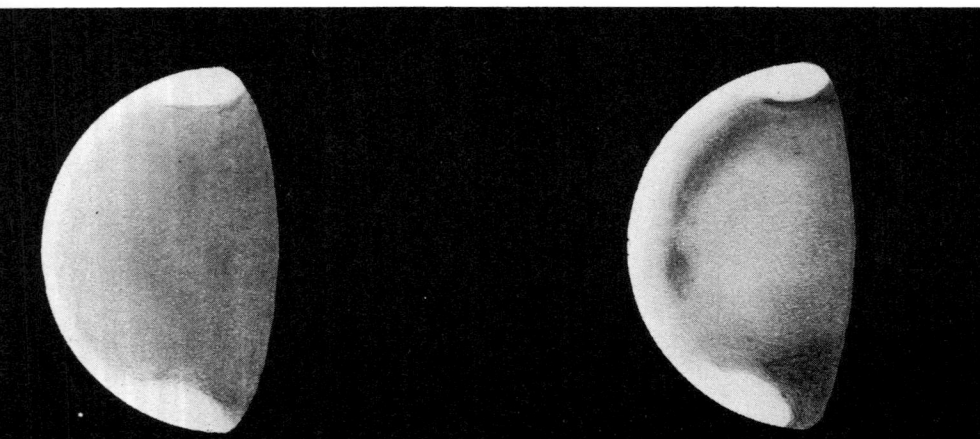

In 1956 I noted a difference in apparent phase when observing in light of different wavelengths through the telescope. It has since been established that the blue light phase is usually smaller than that in red or yellow light (Plate 14). This has a bearing on the accuracy or otherwise of phases measured on drawings, however carefully drawn. The colour sensitivity of observers varies from one observer to another in a marked degree, resulting in an observer who is blue-sensitive, say, making the phase too small.

The phase difference between yellow and red light is minimal, but detail on the disk is not so well seen in red light, and the light areas tend to appear smaller than in blue light. It might be contended that green is the best colour to use, as it is near the middle of the spectrum and easy on the eyes; but, unfortunately, green enhances the light areas, especially the cusp caps (p. 66). Therefore the Wratten 15 (yellow) filter with passband at about 5,500A and downwards into the red has been adopted as the standard wavelength to observe the phases of Venus. This has been used for some years by the Mercury and Venus Section of the British Astronomical Association, and a marked decrease in the scatter of plotted phase measurements from drawings has resulted. One cannot too strongly recommend the use of the Wratten yellow filter, or something similar, for observation of the phases of Venus. Details of the methods of using filters will be found on page 27.

Phase observations are usually plotted in percentages against dates or days. Usually an anomaly is shown, because the date of observed dichotomy very rarely falls on the calculated date. There is also a tendency to underestimate the phase at large values, and this must be allowed for.

The shading of the terminator will appear heavily marked in blue light, but narrower in red or yellow. The intensity of shading will be seen to vary from day to day, probably owing to large air movements in the atmosphere of Venus.

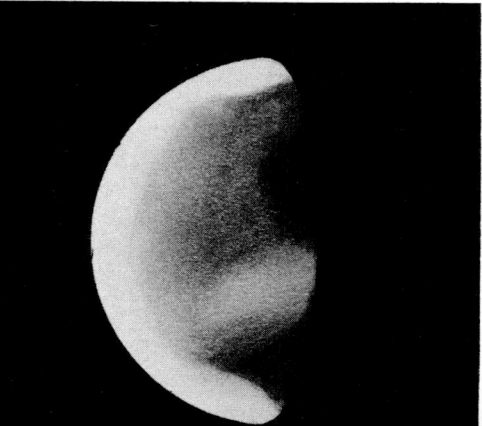

14 Three drawings of Venus 1976 December 28, with an 8½in (216mm) reflector, showing, left to right, the images in red, yellow and blue light (*Richard J. McKim*)

The terminator will also take on irregularities, with 'bays' of light areas invading the dark and 'peninsulas' of dark patches in the light. The light areas are more difficult to see than the dark patches, so it is best to record, first, the phase; then to search for dark patches and streaks; and following this, when the eye is fully accommodated to the conditions, to search for the light areas. These will be found to predominate at the cusps of the crescent and in between the dark shadings. They are also sometimes to be seen on the terminator shading. It is thought the light areas are high altitude clouds in the atmosphere of Venus illuminated by sunlight. This is highly likely, because they appear larger in blue light than in yellow. The limb of the planet is also often bright.

A warning: details on the disk of Venus are difficult to see. Venus is an easy object to see in the sky, but a difficult one to observe. One can use too little magnification in the telescope and so render the image too bright to see detail, but too much magnification will kill the contrast and again render the delicate shadings invisible. Each observer will find his own comfortable light level, at which he will be able, after some practice, to see the features under consideration (Plate 14). Occasionally, especially when Venus is small and far away, with the sunlight pouring straight down on its atmosphere and causing very little in the way of shadows, the disk seems to have no detail at all. When the sunlight strikes Venus at a sharper angle as seen from Earth, the shadows are longer and the detail more easily seen.

Now, as mentioned before, we must consider the cusp caps. These show as brightening of the points of the crescent or at the apparent poles of the gibbous-phased planet. It is thought that they are high level clouds near the poles but not necessarily over them. At some times they are obvious and at others they do not appear, perhaps because of the tilt of the axis of rotation of Venus; but more study is needed before any conclusion can be arrived at. They are most clearly seen in green light, and this suggests they are at low temperature, probably frozen. The observer should always check for their appearance when making a drawing.

The most controversial phenomenon to be observed on Venus is the Ashen Light. This is a glow of a coppery hue that tends to complete the circular disk when the planet exhibits a bright crescent. It must not be confused with the purely subjective appearance of a dark disk completing the figure of the crescent phase.

The best way to make sure that the appearance of the Ashen Light is real is to cover the bright crescent with an occulting bar—a piece of metal, or similar suitable material, mounted in the focal plane of the eyepiece so that it presents a sharp edge when viewed through the eyepiece. It can often be fixed to the field stop in the eyepiece body. The occulting bar can be refined by making it a curve of varying radius, so that one can choose a portion of similar shape to the crescent Venus in the field of view.

The bright crescent should be hidden by the bar, and then the suspected Ashen Light checked to see whether it is still visible. Personally, I often find that the illusion of the presence of the Ashen Light is so strong that I am tempted into thinking it is real; but, when the bar is applied, the light disappears. If the light remains visible when the bright crescent is covered one can be reasonably sure it is real.

Another, perhaps more refined, test is to view the planet, with the crescent covered by the bar, through a Wratten 35 (purple) filter. This transmits light visually similar to the true Ashen Light, and the light should remain visible. One caution is needed here: the 35 filter is difficult to use because the eye does not readily adjust itself to the wavelength, and so becomes strained.

Although not of a telescopic nature, there is another matter that cannot be omitted from our consideration. That is the 4-day period of pattern change in the shadings seen on Venus. While the body of the planet rotates in a retrograde sense once every 243 days, the upper atmosphere rotates much more rapidly and shows a repetition of patterns once every 4 days. This is not a constant feature, but should be looked for when comparing drawings of the planet.

One final word is perhaps necessary on the matter of the delicate nature of the shadings and light areas. With a suitable filter on a telescope of reasonable size (say a 6in (150mm) reflector or larger) and a magnification of about 150 diameters the light level is reduced to a comfortable intensity, but the details are very delicate, and one must be warned against thinking one sees something that is not there. So there is the accepted rule that one must see detail during the best moments of seeing, and do so repeatedly, not just once. This is the art of observing.

Observing Mars

Mars is a terrestrial-type planet that has a rocky surface, as has Earth, and is of interest to the observing astronomer for this very reason. On Mars the various surface details visible in the telescope comprise the poles, the dark areas, and the lighter deserts.

It is true that space probes have disclosed views on the surface of the planet and shown the presence of craters similar to those on the Moon; the Earth-bound observer is unable to see these craters, whose patterns of distribution do not follow the same delineations as the dark and lighter areas, nor the fine detail on the surface. It is nevertheless of importance to keep the planet under scrutiny, especially because there are changes in the appearance of the surface, both seasonal and erratic. In addition, occasional pictures made by visiting space probes only supply momentary views at certain times, and, while they provide much more information than Earth-bound telescopes can ever do, they cannot as yet record changes over long periods. Even the US Viking landers can supply only local information in most fields. One can say, therefore, that telescopic observation of the planet is still desirable; indeed, it could be more correctly said that telescopic observation remains imperative.

Unfortunately for continuous observation, Mars approaches Earth only once every two years, the mean synodic period being 779·94 days. A

15 Drawing of Mars 1973 October 5, 22hr 30 min UT, with 10in (254mm) spec and magnification of 300 diameters. Longitude of central meridian 322·8 deg. (*P. B. Doherty*)

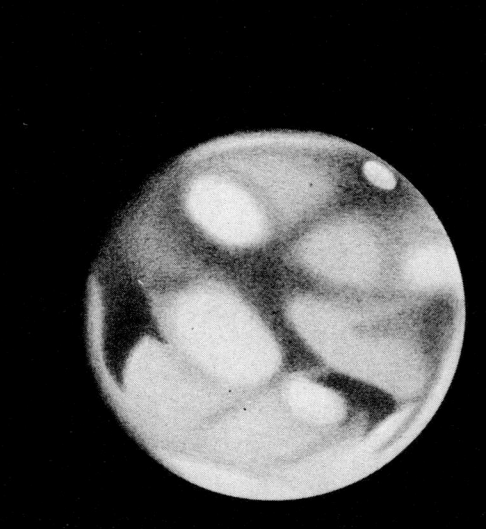

further difficulty arises for northern-hemisphere observers in that, when Mars is closest to Earth, it has a southern declination and so is not well placed for observation.

At its closest approach Mars subtends an angle of 25·7″, and at its greatest distance only 3·5″. This means that Mars is only really worth observing during short periods around the time of apparition when its diameter is not less than 7″.

Owing to the eccentricity of the Martian orbit, the diameter of the planet can reach 25·7″ at the most favourable apparition, but only 13·8″ when the apparition is not so favourable. It is obvious that Mars is most worth observing during its closest approaches, which occur during August and September, when Mars is south of the Equator and best seen from the southern hemisphere. At the time of the best apparitions the south pole of Mars is turned towards the Earth, while at the less favourable times the north pole is presented. Table 7 lists oppositions.

Table 7

Oppositions AD 1971–99

Date of opposition			Apparent diameter (sec)	Declination (deg)
1971	August 10	7 hr	24·9	−22
1973	October 25	3 hr	21·5	+10
1975	December 15	4 hr	16·6	+26
1978	January 22	0 hr	14·3	+24
1980	February 25	6 hr	13·8	+14
1982	March 31	10 hr	14·7	− 1
1984	May 11	9 hr	17·6	−18
1986	July 10	5 hr	22·1	−28
1988	September 28	3 hr	23·8	−2
1990	November 27	20 hr	17·8	+23
1993	January 7	23 hr	14·0	+27
1995	February 12	2 hr	13·9	+18
1997	March 17	8 hr	14·2	+ 5
1999	April 24	18 hr	16·2	−11

A telescope of 3–4in (75–100mm) diameter will show during a favourable apparition the poles, as white caps, and some of the grosser markings. At other times only the polar caps will be clear with an instrument of this size. For more serious work a larger instrument is essential, and a reflector of 8 or 10in (200 or 250mm) is considered the minimum. Do not be put off by hazy skies, for mist can suppress the glare and make detail more easily seen in steady air.

Magnification to be used is a matter of local conditions, and the aim should be to obtain a clear picture and not to crowd on magnification for the sake of enlarging the image. If seeing is good, one could suggest as a guide a magnification of about forty to the inch of objective or mirror (1·5 to the millimetre) for larger instruments, but this again is a matter of local atmospheric conditions. A poor image is poor whether magnified or not. In most cases a magnification of from 200 to 400 should meet the demands of the observer, and give the clearest image (Plate 15).

Since the observation of Mars requires comparatively large magnification, it is essential that the telescope mounting be rigid and free from shake, and a clock drive to hold the image still in the field of view is an advantage.

When observing Mars, it is a good plan to look steadily at the planet for a time to let the eye accustom itself to the conditions, and then by trial and error to find the best magnification to use to get the clearest picture of the planet. Do not focus on the hazy patches on the disk, but first focus on the limb of the planet to be sure you have the clearest setting of the eyepiece. Then, after so doing, take note of the more hazy details.

Concentrate first on the white polar cap or caps, and, on a standard size disk of 2in (50mm) diameter, draw them in (anything smaller may cause overcrowding of the drawing with detail at a later stage, while anything larger tends to make it awkward to space out the details proportionally). Be very careful to get the proportions right, for if the poles are not accurately delineated the whole picture will be out of character.

Fig 14 The surface of Mars. Areas suspected of change are numbered as follows:

1 Aethiopis	5 Solis Lacus	9 Daedalia-Claritas
2 Cyclopia	6 Araxes	10 Hellas
3 Thoth-Nepenthes	7 Oxia Palus	11 Syrtis Major
4 Pandorae	8 Ganges	

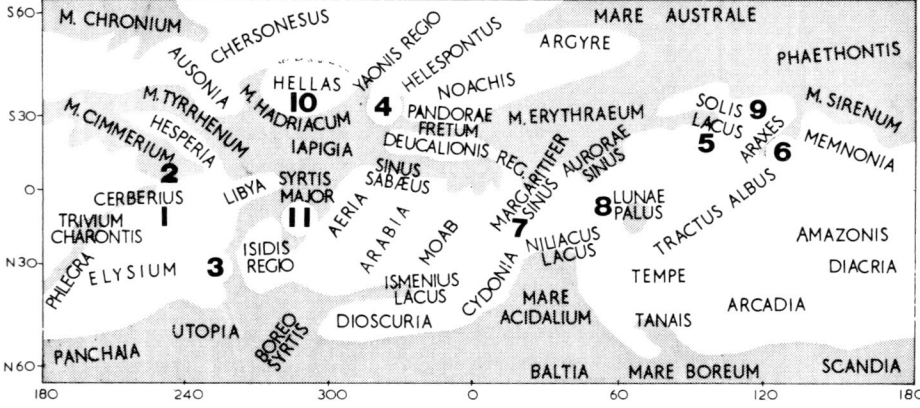

Having recorded the polar caps and obtained some sense of general proportion for the drawing, start examining the disk for dark shadings or markings. A little patience and careful examination will here repay the observer, who should only record what he sees well; glimpsed or imagined details, subconsciously recalled from previous knowledge of the topography of Mars, should be ignored. It is surprising what can be seen if one already knows what one is supposed to see when working at the limits of the telescope's capability and one's optic vision.

It is a good plan to sketch in the outlines of any dark markings. Afterwards you can shade them in as accurately as possible to portray the various intensities of shading.

Orientation of the drawing should be shown, and can be obtained from the line of drift of the planet in the field of view, considering that Mars will move towards the west if the telescope is kept still. This can be accomplished by stopping the driving clock, if one is fitted, for a few moments and noting the direction in which the planet moves. If using an instrument not equipped with a clock drive, move the telescope in advance of the planet and wait for any shake to die down, during which time the planet will drift into view. Make the observation, and then move the telescope forward again and repeat the process. Do not attempt to draw any fine detail as seen through an unsteady instrument, nor any detail you have poorly grasped; it will lead only to confusion.

Modern observation of Mars is concerned with patrolling for changes in the appearance of the features visible. The polar caps naturally fall into this category, but there are many other areas of the Martian globe that are worth checking for variations in shape and intensity of albedo. A map of Mars with these particular areas numbered is reproduced here (Fig 14), but is not intended as a reference map for fine detail.

For the reason already stated, I find it best to observe Mars without making any reference to the map or the longitude of the centre of the disk presented in the telescope before going to the observatory. This keeps my mind clear of bias. The better way is to draw Mars at the telescope, and then, and only then, when the observation is completed, to refer to the map for confirmation or otherwise of the results. For this purpose, a good detailed map, such as that produced by the International Astronomical Union, and provided for members of the Mars Section of the British Astronomical Association and other bodies, is recommended.

In this connection it is necessary to know the longitude of the centre of the presented disk of the planet. This is published for each day in the *Ephemeris* (as at oohr UT each day), when Mars is available for observation. It will be known that Mars rotates on its axis once in 24 hr 37 min 23 sec, and the longitude of the centre of the disk can be calculated from the figure given in the *Ephemeris*, duly corrected by taking into account the amount of rotation during the period from oo hr UT to the time of the observation.

A table of the change in longitude of the central meridian of Mars (in terrestrial time) is given below to assist observers in establishing the correct longitude for their drawings.

Table 8

Change of Martian Longitude

min	o hr	1 hr	2 hr	3 hr	4 hr﹙	5 hr	6 hr
	°	°	°	°	°	°	°
o	0·00	14·62	29·24	43·86	58·48	73·10	87·72
5	1·22	15·84	30·46	45·08	59·70	74·32	88·94
10	2·44	17·06	31·68	46·30	60·92	75·54	90·16
15	3·66	18·28	32·90	47·52	62·14	76·76	91·38
20	4·87	19·49	34·11	48·74	63·36	77·97	92·60
25	6·09	20·71	35·33	49·95	64·57	79·19	93·81
30	7·31	21·93	36·55	51·17	65·79	80·41	95·03
35	8·53	23·15	37·77	52·39	67·01	81·63	96·25
40	9·75	24·37	38·99	53·61	68·23	82·85	97·47
45	10·97	25·59	40·21	54·83	69·45	84·07	98·69
50	12·18	26·80	41·42	56·05	70·67	85·29	99·91
55	13·40	28·02	42·64	57·26	71·88	86·50	101·13

1 min=0·24   2 min=0·49   3 min=0·73   4 min=0·97   5 min=1·22

It naturally follows that Mars takes 37 min 23 sec longer to rotate once than the Earth does. Thus, if one observes an object near the centre of the disk of Mars early one night, then the same feature can be seen again the following night 37 min 23 sec later. In practice the seconds can be ignored. By taking advantage of this you can obtain a series of observations of any specific Martian feature and achieve a detailed study.

Frequently clouds in the Martian atmosphere cover some areas, and dust storms such as those of 1971 and 1973 almost completely obliterate detail over the whole of the disk. The positions, duration and movements of any clouds seen should be noted on a special separate drawing. The Wratten 47 (blue) filter will be found a great help in allowing you to see these clouds and plot their positions on the disk. From these observations the ground speed of the Martian air currents can be calculated.

It would be well here to caution the new observer regarding the apparent distortion of the disk, which is at times quite evident. The disk outline is not always an accurate circle because, although in an orbit outside that of the Earth, Mars does show some phase early and late in an opposition.

The planet is then apparently small, but the phase defection can reach a maximum of 15%.

Prepared disks taken to the telescope can be of assistance here, but printed disks of circular form can be adjusted at the telescope by drawing in the apparent terminator, before attempting to record other features. Care must be exercised because, as we have said, the image of the planet is small at this stage.

Observing the Asteroids

The minor planets, or asteroids as they are more often called, are often described as having orbits between those of Mars and Jupiter. This is largely true, but an asteroid such as Icarus passes so near the Sun that it travels for a part of its orbit within that of Mercury, while at the other extreme Hidalgo travels beyond the orbit of Jupiter. Many of these little bodies have greatly inclined orbits so that they can be seen far from the plane of the ecliptic.

There are two types of observation which owners of moderately sized telescopes can carry out: these are observations of apparent magnitude (or apparent brightness) and observations of apparent position. Observations of magnitude can be carried out in the same manner as that suggested for variable stars (see page 99). In this type of work the apparent brightness of the asteroid is compared with that of nearby stars.

To accomplish this it is first necessary to ascertain the position of the asteroid from a good ephemeris and to plot this on a star map. Make your own map of the particular part of the sky to a scale large enough to be seen easily by the dim red light used at the telescope. Compare the map with the sky and mark in the position of any star suspected of being an asteroid by reason of its not being on the original star map. It will suffice to differentiate between stars and asteroids by marking the former as black dots and the latter as circles. After the lapse of some few hours, the asteroid you are looking at will be seen to have moved among the fixed stars. This establishes its identity.

Table 9

Interesting Asteroids

Number	Name	Year of discovery	Diameter (km)*	Sidereal period (years)	Mag†
1	Ceres	1801	1,003	4·60	7·4
2	Pallas	1802	608	4·61	7·8
3	Juno	1804	247	4·36	9·2
4	Vesta	1807	538	3·63	6·8
5	Astraea	1845	117	4·14	—
6	Hebe	1847	201	3·78	8·4
7	Iris	1847	209	3·68	8·2
8	Flora	1847	151	3·27	8·5
9	Metis	1848	151	3·69	9·2
10	Hygiea	1849	450	5·60	10·0
11	Parthenope	1850	150	3·84	10·2
12	Victoria	1850	126	3·56	—
13	Egeria	1850	224	4·14	—
14	Irene	1851	158	4·16	9·8
15	Eunomia	1851	272	4·30	8·5
16	Psyche	1852	250	4·99	10·2
18	Melpomene	1852	150	3·48	9·2
19	Fortuna	1852	215	3·82	—
20	Massalia	1852	131	3·74	9·3
22	Kalliope (Calliope)	1852	177	4·96	10·6
27	Euterpe	1853	108	3·60	10·4
29	Amphitrite	1854	195	4·08	9·8
30	Urania	1854	91	3·64	10·5
39	Laetitia	1856	163	4·60	10·3
40	Harmonia	1856	100	3·41	10·5
44	Nysa	1857	82	3·77	9·9
51	Nemausa	1858	151	3·64	—
63	Ausonia	1861	91	3·70	10·5
192	Nausicaa	1879	94	3·72	9·9
324	Bamberga	1892	246	4·40	9·1
349	Dembrowska	1892	144	5·00	10·8
511	Davida	1903	323	5·72	10·3
532	Herculina	1904	150	4·61	—
433	Eros	1898	23	1·76	8·8
1566	Icarus	1949	1	1·12	13·0

*The diameters are those derived from polarimetric and radiometric methods, and published in 1977.

†These figures are the maximum stellar magnitudes of certain asteroids as seen from Earth during the last few years. Most asteroids will be observed to be fainter than the figures quoted, but these will give some idea of the possibilities of catching sight of one of these little bodies.

Now look for stars of similar brightness or magnitude and make the comparison. For this purpose use a low magnification eyepiece with a large field of view. Some asteroids can be observed even with binoculars in this way, especially during their close approaches, when they are brighter than the mean value.

The apparent magnitude will be seen to change over a period owing to the changing distances between the asteroid, the Sun and Earth. The brightest appearance will obviously occur when an asteroid is near both the Earth and the Sun (see Appendix 1).

Observations of position by visual sighting and then plotting are not really accurate enough for modern requirements, but they can afford some valuable experience and enjoyment to the observer. Take the position from a good ephemeris and star map, as previously mentioned, and then, having established the identity of the asteroid, mark in its position from night to night. It will be seen to move against the background star pattern; if near a star, the asteroid can be seen to move during one night's observation only. The effects of precession must be taken into account when preparing the star map (see page 35). Bring the values to the same date as the ephemeris of the asteroid by applying the required corrections to the star positions.

It is sometimes possible to follow an asteroid for a long period of time, and in this case it may become necessary to draw extension maps of the sky for comparing its position and magnitude in relation to the stars. Table 9 lists some of the asteroids.

Observing Jupiter

From the smallest planets in the Solar System, the asteroids, we now turn our attention to the largest planet—Jupiter. This planet is the largest body in the Solar System, excepting the Sun itself, and has an equatorial diameter of 142,200km.

At the distance of 4·2AU (astronomical units) when it is due south at midnight, it is a very obvious object in the sky, subtending a diameter of some 47″. It is thus a comparatively large image in the eyepiece, and detail can be easily distinguished. Jupiter has on this account been referred to as the happy hunting ground of the amateur astronomer. It is indeed a magnificent sight.

A very small telescope will show the four major satellites, and a telescope of as little as 2in (50mm) diameter will show evidence of the belts running round the body of the planet and extending along parallels of Jovian latitude. The serious student of Jupiter will need something larger than a 2in (50mm) telescope, however, and a reflector of 6in (150mm) with magnification of 150 or more is recommended as about the lower limit for proper observation. A 4in (100mm) refractor is also suitable. The larger the telescope, of course, the better; more accurate and more detailed information will be obtained with larger instruments.

It is a good exercise sometimes to draw the detail seen on the disk. Because of the rapid rotation of Jupiter it is advised to start drawing at the preceding limb, as detail will disappear here while the drawing is being made; the following limb will be bringing detail onto the disk at the same time, and so the drawing can be finished here. In all, a time of no more than 20 min should be allowed for the complete drawing because of the effects of the rapid rotation.

For the serious observer, the first thing to do is to get to know the names of the various belts and zones for the sake of accurate reference. (See Fig 15.)

The rotation of Jupiter is more rapid near its equator than towards the poles, becoming progressively slower with higher latitudes. Some system of reference is necessary, for it is obviously impossible to assign a standard rotation speed to each and every belt and zone. Recourse is therefore made to two standard systems, as follows:

System I refers to features in the Equatorial Zone, the southern edge of the North Equatorial Belt, and the northern edge of the South Equatorial Belt. It is allotted a standard period of rotation of 9 hr 50 min 30 sec.
System II refers to all other latitudes and is allotted a rotation period of 9 hr 55 min 40·65 sec.

We thus establish two independent systems of reckoning Jovian longitude. They are referred to an imaginary line running due north and south down the centre of the apparent disk of Jupiter, and the longitude of any feature on the centre of the disk at any particular instant is the longitude of the central meridian of Jupiter at that instant in time.

To establish the longitude of any particular feature, note the time of its crossing the central meridian to the nearest minute. The longitude of the central meridian is ascertained from the *Ephemeris* at a particular date and time; adjustment must then be made for the time interval between the *Ephemeris* figure and the time of the observation. Readers

THE BELTS AND ZONES

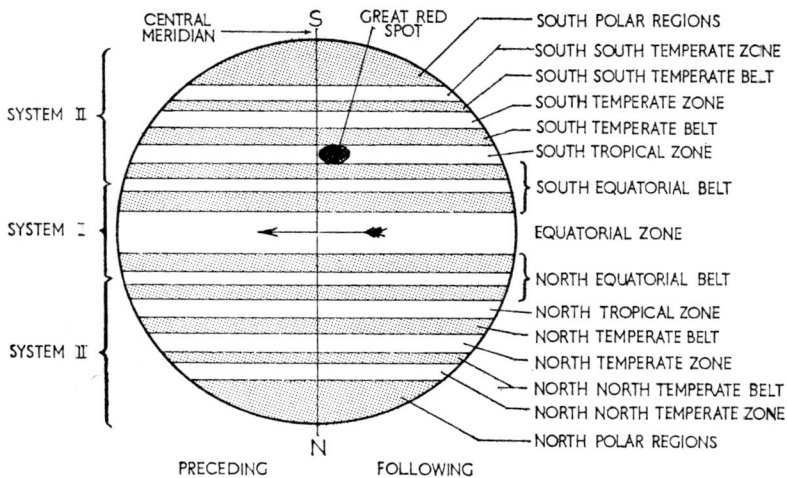

Fig 15 The belts and zones of Jupiter

must refer to the *Ephemeris* for the datum longitude, but Table 10 gives the degrees of change in longitude in intervals of mean time for Systems I and II.

These are the amounts to be added to or subtracted from the datum figures expressed in degrees; if one observes before the time quoted against any particular day in the *Ephemeris*, the change in longitude must be *subtracted* from the datum longitude, and if one observes after the datum time, the change must be added to the datum value.

The rapid rotation of Jupiter accounts for the marked oblateness of the planet; no one seeing Jupiter in a telescope can fail to note this 'flattening' at the poles and 'bulging' at the equator. This distortion of the disk must

Table 10   Jupiter

CHANGE OF LONGITUDE IN INTERVALS OF MEAN TIME
System I: Period 9 hr 50 min 30 sec

hr	°	hr	°	min	°	min	°	min	°
1	36·6	6	219·5	10	6·1	1	0·6	6	3·7
2	73·2	7	256·1	20	12·2	2	1·2	7	4·3
3	109·7	8	292·7	30	18·3	3	1·8	8	4·9
4	146·3	9	329·2	40	24·4	4	2·4	9	5·5
5	182·9	10	5·8	50	30·5	5	3·0	10	6·1

CHANGE OF LONGITUDE IN INTERVALS OF MEAN TIME

System II: Period 9 hr 55 min 40·65 sec

hr	°	hr	°	min	°	min	°	min	°
1	36·3	6	217·6	10	6·0	1	0·6	6	3·6
2	72·5	7	253·8	20	12·1	2	1·2	7	4·2
3	108·8	8	290·1	30	18·1	3	1·8	8	4·8
4	145·1	9	326·4	40	24·2	4	2·4	9	5·4
5	181·3	10	2·6	50	30·2	5	3·0	10	6·0

be taken into account when drawing detail seen on the planet, particularly for the whole-disk drawings mentioned above although it must also be remembered when drawing details of portions of the disk, for the shape of the limb will not follow the circumference of a circle.

A suitable size of disk measures 1·96 × 2·1in (49·5 × 53mm). This can be drawn on a piece of card and then cut out for use as a template for making Jupiter disks. If something a little larger is thought to be better, then the diameters may be made 2·32 × 2·48in (59mm × 63mm), but do not make the disk too large or there may be trouble in getting the drawing in proportion. (See Plate 16.)

While drawing detail is an interesting pastime, useful for the description and recognition of particular features, the main business of the serious observer is the recording of transits over the central meridian of specific features for the purpose of obtaining their longitudes. When a number of transit times have been converted to longitude, they are plotted against date on a graph in either System I or System II. Thus some knowledge of the various features is gained, and their behaviour can be followed.

On the practical observational side, there are two suggestions worthy of mention. First, let the eye get used to the conditions before attempting to record features. Given time, the eye will accommodate itself to the light-level and detail will begin to appear more clearly. A novice often finds he can see very little on the disk of Jupiter, whereas a practised observer will see much more. Second, do not attempt to use too high a power of magnification, somewhere about 180 to 200 diameters being quite sufficient. To crowd on magnification to too high a value will only result in poor definition and loss of contrast.

One salient feature that cannot be missed is the well known Red Spot. When on the disk, it sometimes appears very clearly red, at others pink, and on some other occasions will almost disappear. Its longitude also varies considerably. The behaviour of objects in its vicinity is of much interest, and observations are most welcome (Plate 16).

This brings us to the matter of colour of the various features on the disk. Together with A. W. Heath, I have done some research on this matter, and we have shown that the ruddy and grey colours are real, and not contrast effects. Records of changing colour, which can be of value, are

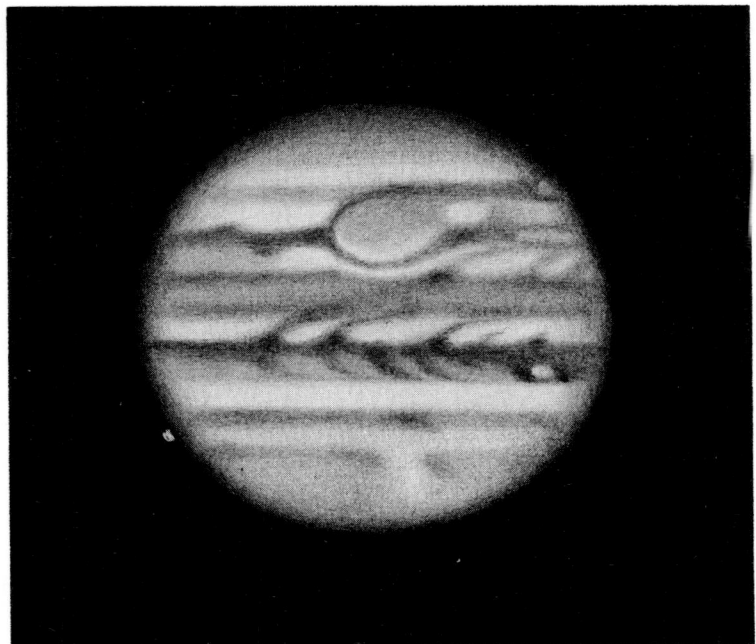

16 Drawing of Jupiter 1976 December 28, 18 hr 44 min UT, with 16·5in (419mm) spec, X300 magnification. Note the Red Spot and Satellite II entering into occultation. Longitudes: I 26·4°, II 43·3°. Compare with Plate 7 (*P. B. Doherty*)

obtainable by making intensity estimates on a scale of zero to ten, zero representing the brightest part of the Jovian disk and ten total blackness. Estimates are made in red, white and blue light, using colour filters. The intensities are then plotted on graphs against dates, when variations will show up.

Jupiter, being predominantly of a ruddy colour in the belts, exhibits detail well with a light blue filter, which suppresses the red light and renders the ruddy features black, making them easier to see. I have used the Wratten 44A for some time to good effect.

Interesting observations can be made of the behaviour of the four major satellites. These can sometimes be seen in transit across the face of Jupiter, when they will cast shadows on the planet either east or west of the satellite, according to the relative position of Jupiter and Earth in their orbits. The satellites are also often eclipsed by Jupiter's own shadow, and can be followed through these events with quite a small instrument. They are, too, often occulted by Jupiter.

When the Earth passes through the plane of the orbits of the satellites, they can be seen to eclipse and occult one another. This happens once every six years, and the timing of any of these events is of value in checking the orbits of the moons.

When passing in front of Jupiter, satellites III and IV are generally dark and I is faint grey; II is very difficult to detect against the bright background of the planet's disk.

Observers trying to coordinate visual features with the radio pulses from Jupiter may be interested to know that a third system of rotation period is now used in this connection, 9 hr 55 min 29·71 sec. The outbursts are affected by satellite I (Io), whose synodic period of revolution round Jupiter is 1 day 18 hr 28 min 35·9 sec.

Observing Saturn

If Jupiter is a magnificent sight in the telescope, Saturn must be rated the loveliest planet. It has been described as the most beautiful inanimate body in the Solar System, and rightly so. Saturn has a globe with belts and zones, as does Jupiter, though these are not so active nor so prominent and impressive, but it is surrounded by its system of rings that add so much to the beauty of its appearance.

Saturn will bear greater magnification than Jupiter, and it is therefore usual to scrutinise the planet and ring systems with a higher magnification, say something in the region of 200 diameters and upwards, always supposing atmospheric conditions will permit the use of short-focal-length eyepieces. As with Jupiter, a light blue filter will help to bring out detail.

Saturn has been neglected by many observers because it seems to lack the dramatic upheavals and disturbances of the belts of Jupiter, and at first glance the rings seem more or less static in appearance. Detailed examination will show this not to be so, however, and the planet requires constant and careful observation.

Comparatively little is known about the ring systems, though they are presumed to be composed of rocks of up to 3ft (1m) in diameter with ice crystals included in their make-up. The period of rotation of the ring systems is more or less a matter of conjecture at present, and telescopic search for irregularities that might lead to some method of measurement is needed. Observation of the movement of a dark or light area on a ring might result in the establishment of a proper rotation period. This will, of course, need a telescope of considerable aperture and an observer of some experience.

Work is also needed on the establishment of the rotation period of various parts of the globe of Saturn. The equatorial zone is known to rotate once in 10 hr 14 min, but Saturn is a gaseous globe like Jupiter, and this figure is not applicable to the whole of the planet (see Appendix 2).

In the main the spots that do appear are usually few and short lasting. Examination of both rings and globe is therefore necessary on each and every occasion of observation.

A caution would not be out of place here. It is not always easy to decide what one really sees under high magnification and perhaps disturbed air conditions. The art of observation lies in deciding what is real and what is illusory. All irregular appearances, once established as real, should be recorded accurately against time: there is always the chance that another observer may be able either to confirm the sighting or to deny its reality.

The sighting of anomalies is connected with the quality of the seeing conditions very closely. A useful test of these conditions is to look for the Cassini Division between rings A (the outermost) and B, which is brighter than A; if the Division cannot be seen, do not attempt serious observation. The dusky crepe ring known as ring C is the innermost one visible in normal-sized telescopes, and can often be seen as a dusky streak against the bright globe of the planet. There are other rings both inside and outside the main system, but we are not concerned with them, since they are beyond the capabilities of the normal instrument and require photographic techniques in very clear air for their detection.

To return to the globe of Saturn, the rotation period of the equatorial zone, given as 10 hr 14 min, cannot be used as a method of establishing longitude, since no first meridian has been assigned to Saturn; by timing the rotations we can, however, use an arbitrary system for recording the position of any feature in longitude with reference to another feature (see Appendix 2). We can also use the known rotation period of the equatorial zone to find the day and time at which we can re-observe the same part of the zone. Tables 11 and 12 serve this purpose and allow the establishment of change in longitude over given periods of time.

Table 11

Rotation of Saturn

No of rotations	day	hr	Period min	To observe the same part of the equatorial zone			
1		10	14	after 1 day observe	3 hr	32 min	earlier
2		20	28	after 2 days	3	10	later
3	1	6	42	after 3	—	22	earlier
4	1	16	56	after 4	there will be no repeat appearance		
5	2	3	10				
10	4	6	20	after 5	2	38	later
15	6	9	30	after 6	—	44	earlier
20	8	12	40	after 7	4	16	earlier
25	10	15	50	after 8	2	26	later
30	12	19	00	after 9	1	6	earlier
35	14	22	10	after 10	4	38	earlier
40	17	1	20	after 20	—	58	later
45	19	4	30				
50	21	7	40				

Table 12

Change of Longitude, Adopting Rotation Period of 10 hr 14 min

Number of days elapsed	Change in longitude (°)	Number of days elapsed	Change in longitude (°)
50	95·0	4	137·2
40	292·0	3	12·9
30	129·0	2	248·6
20	326·0	1	124·3
15	64·5	*Time elapsed*	
10	163·0	12 hr	62·2
9	38·7	6	211·1
8	274·4	3	105·5
7	150·1	2	70·4
6	25·8	1	35·2
5	261·5	$\frac{1}{2}$	17·6

For smaller time intervals divide accordingly

The arbitrary longitude of any feature is obtained by timing transits over the central meridian in the same manner as for Jupiter.

Determining latitude on the globe of Saturn is a very different matter. If Saturn maintained a constant tilt relative to the line of sight from Earth, it would be simple, but the planet presents different angles of inclination with respect to Earth according to orbital position. The tilt is quoted in the *Ephemeris* but, since Saturn is a globe, the adjustment is not simply a matter of elevation or depression of the centre of the disk, for the lateral displacement varies in a decreasing way as one approaches the apparent limb, either north or south.

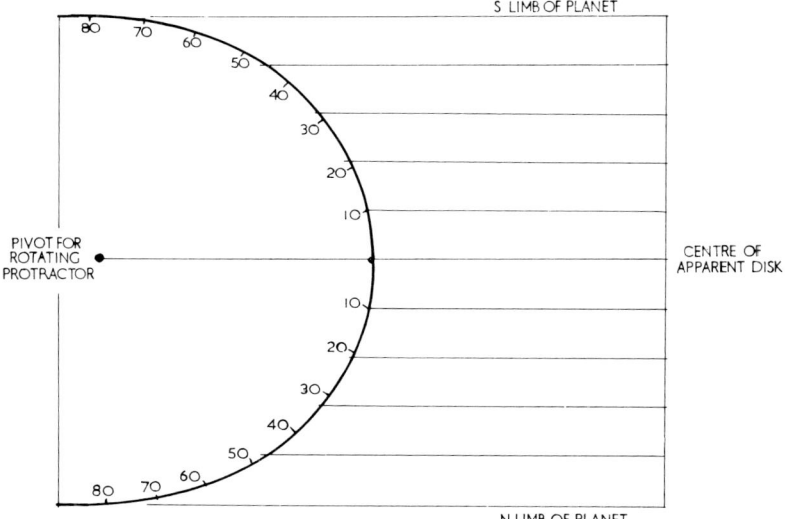

Fig 16 Saturnicentric latitude device

I have produced a device for reading off the true latitude against a scale of ten intervals on a line running north-south on the apparent disk, as seen in the telescope. The centre of the disk is shown, with five intervals north and five intervals south of it, to each limb. Against this scale a rotatable protractor is set to read the tilt, and the true latitude can be read off against any interval or fraction of it. The protractor is not a true semi-circle, but is shaped with a 10 per cent increase in radius on the equator in comparison with the polar diameter. This takes care of any distortion due to the ellipticity of the globe (Fig 16).

Records of the change in latitude of the belts and zones are of particular value. Micrometers are often used for these measurements, but I am of the opinion that equally accurate results can be obtained by using the device described above. It is not difficult to construct, using stiff card, and the protractor can be pivoted on a ladies' dress press-stud. A little practice will enable reasonably accurate measures of Saturnicentric latitudes to be made, but care will be needed to prevent the observer being misled by the presence of the rings in the field of view.

The changing tilt is not confined only to the globe of Saturn, of course, but applies also to the ring systems, which are coplanar with the equator of Saturn. This causes the rings to appear at different angles, so that they look wide open at one part of the orbit, partly closed at others, and sometimes seem to disappear when 'edge on' to observers on Earth. They can also lie edge on to the Sun, so causing diminution of their brightness. (See Plates 17 and 18.)

17 *Above*, drawing of Saturn 1973 November 11, 01 hr 20 min to 01 hr 30 min UT, 10in (254mm) spec, X250. Note rings are near widest opening. *Below*, drawing of Saturn 1976 December 28, 23 hr 42 min UT, 16·5in (419mm) spec, X300. Compare with Plate 8 (*P. B. Doherty*)

18 Drawing of Saturn 1966 October 25, 20 hr 20 min to 20 hr 30 min, 8·5in (216mm) spec, X274. Note satellite on following ansae (*P. B. Doherty*)

Eight of the ten satellites orbit close to the plane of Saturn's rotation. The brighter of these can be seen in a reasonably large telescope, but it will be found best to hide the bright planet with an occulting bar or similar device. The satellites are not easy to distinguish from faint stars, and their positions relative to Saturn should be ascertained from the *Ephemeris* before a search is made, to avoid misidentification. Titan is brighter than the others, and usually apparent; Rhea is only a little less bright than Titan; but the others are fainter, with Phoebe too faint for ordinary telescopes altogether. They are best seen when the rings are apparently closed and so less dazzling.

Intensity estimates of the various features of planet and rings are not only of great interest but can also be useful as a means of extending our knowledge. A. W. Heath, C. R. Munford and I have recently done some work in this matter (using red and blue filters) similar to the previously mentioned observations on Jupiter, with ring B regarded as intensity 1 and black darkness represented by 10. We have found the southern equatorial belt predominantly red, some redness in the equatorial zone, and greyness in the polar regions and ring A.

For drawing Saturn, templates are useful, and specimens are shown in

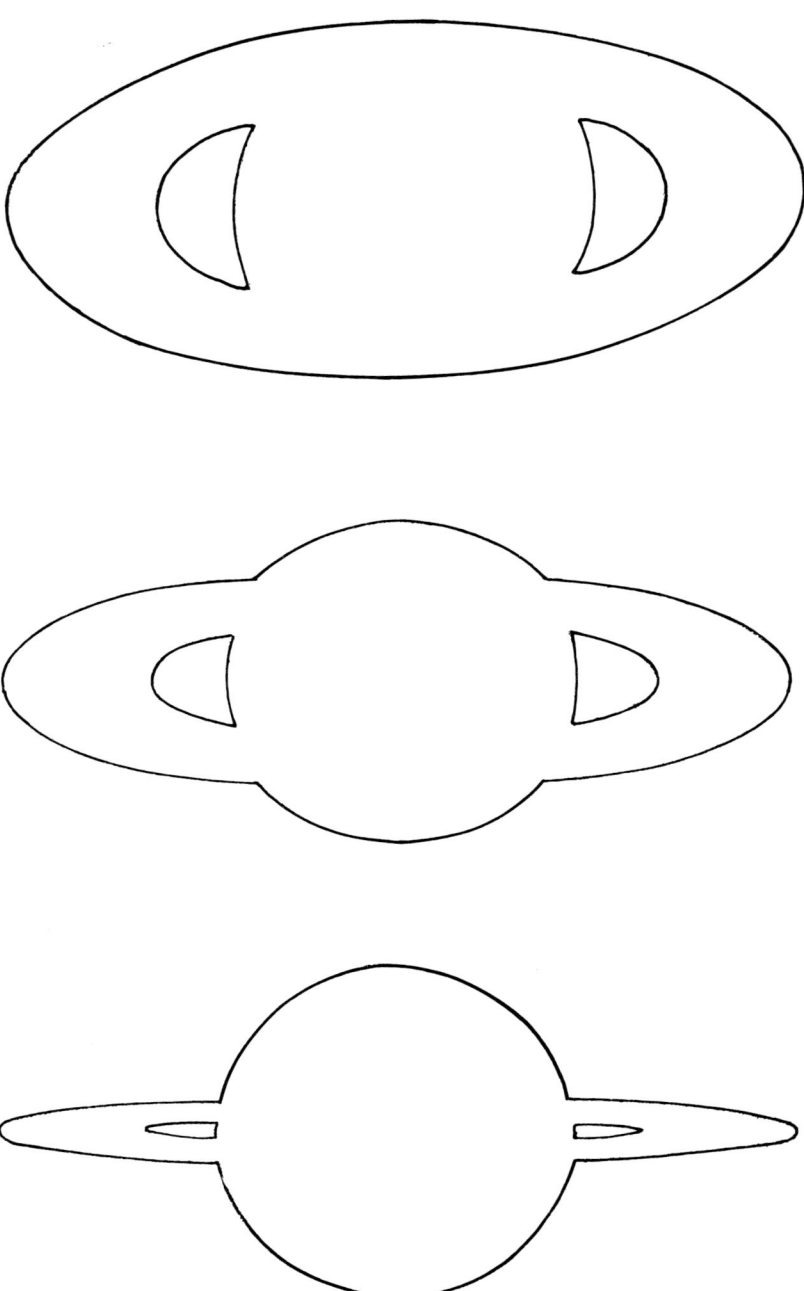

Fig 17 Three specimen templates for drawing Saturn

Fig 17. They are drawn on card to a diameter of about 4in (100mm) across the major axis of the rings, and cut out.

Watch should be maintained for the occultation of stars by Saturn or its rings. Sometimes a star can be seen through the rings.

The appearances of the shadow of the rings on the ball of the planet, and, conversely, the shadows of the ball on the rings, are worth watching. Anomalous appearances are not uncommon, and a regular watch is advised. When these anomalies are seen, a drawing is the best way of describing their appearances, as it will convey much more accurately than a photograph the shape and position of such phenomena.

From the foregoing, it will be realised that, despite first impressions to the contrary, there is much to be done in the way of observation of Saturn. It is worth while to check the appearance of Saturn's globe and ring systems for any peculiar appearance on each and every occasion when observation is possible. One never knows when something unusual may occur.

Observing Uranus, Neptune and Pluto

Uranus

This planet, although in reality about half the diameter of Saturn, is so far from the Sun—and hence from the Earth—that it subtends an angle of only 3·5″ in the telescope. Large magnification is therefore necessary if any detail is to be seen on its surface, and instruments of wide aperture are needed.

The large inclination of the axis of rotation, amounting to some 98°, causes the planet not only to appear to rotate in a retrograde manner, but also to present its poles toward Earth at certain points in its orbit. Uranus was seen to be 'pole on' in 1946 and will be so again in 1985. Between the pole-on appearances a series of bright belts can be seen crossing the disk at various distances from the centre, according to the latitude of the centre of the disk so presented.

Watch should be kept for any apparent detail that may be seen with a view to establishing an accurate rotation period. This has, in the past, been accepted as 10hr 49 min, but recent observations suggest a period between 17 and 22 hr.

Some variation in the apparent brightness of Uranus has been reported, and this is a point that needs looking into; it may be checked by reference to suitable stars near Uranus in the manner described on page 99. This work may be done with binoculars, since the mean opposition magnitude is as bright as 5·5. But it must be realised that for work on the disk or the satellites large telescopes are necessary. The large numbers of minor particles orbiting Uranus, and often loosely referred to as 'rings', are invisible from Earth.

Neptune

Neptune presents a disk just under two-thirds the diameter of Uranus, and so requires large magnification and a large telescope to make observations of any real value. The disk is slightly bluish in colour, and presents a diameter of some 2·3″. Under reasonable conditions a good 3in (75mm) diameter telescope should show this, but the detection of markings is another matter, and needs a large instrument and good atmosphere, with a magnification in the region of 300. The equatorial region has been reported as bright with the poles dusky, but this was with the Mount Wilson 60in (1524mm) telescope and a magnification of 1,000 times.

Satisfactory observation of Neptune is really beyond the capabilities of the average telescope. Its position in the sky, when plotted against an atlas, is possible to establish, however, and with magnitude 7·8 Neptune is not difficult to see.

Pluto

Here the situation is more difficult than for Neptune, for the mean opposition magnitude of Pluto is 14·9. However, Pluto is moving in closer to the Sun and therefore appears brighter than the mean opposition value; the current (1978) magnitude is 14 and the planet can therefore be seen with a telescope of 10in (250mm) diameter. Pluto's diameter at mean opposition distance is only about 0·2″, and for observing the disk a large instrument is necessary. Observation of any surface detail whatsoever is beyond the capabilities of any but the largest telescopes.

Observing Comets

The primary instruction in the recipe for jugged hare is: 'First catch your hare.' Similarly, the primary operation in comet observation is: 'First catch your comet.'

There are two ways of doing this. Firstly, one can plot on a star map the position of a comet whose orbit is known and make a search around the derived position. Secondly, one can search for a hitherto-unobserved or new comet.

Taking the latter possibility first, it is best to search where comets are most likely to be found; i.e., in the western sky just after sunset or in the eastern sky before sunrise. This is because, while comets may appear anywhere in the sky, most comets are discovered in the region inside the orbit of Mars, and they are brightest when nearest the Sun.

A telescope of quite moderate size, with a low power of magnification and large field of view, is adequate for the search. Even an object glass of only 3in (75mm) can be used, although something larger will have obvious advantages.

The usual method is to sweep by traversing the instrument over a definite area of sky which the observer should get to know intimately. A good star map like the *Atlas Coeli*, which can be bought in the field edition on cards, is essential, while *Norton's Star Atlas* can be used for brighter objects down to the limiting magnitude of naked-eye stars. All nebulae and clusters should be noted, so as not to confuse them with a possible comet. If a star cluster is seen to resemble a comet, use a larger magnification on the specific object to check its real nature; faint star clusters and nebulae can look very like comets at low powers of magnification. If an unidentified object is seen, draw a map of the area and wait a few hours, then compare the map with the sky again. If a comet has been sighted, its position relative to the stars will have changed in the interval.

The position of the comet in Right Ascension and declination should be established by reference to the star map, and not estimated by using the setting circles. The circles are not accurate enough for this purpose.

Remember also that a comet does not necessarily show a tail. In fact many comets at discovery have no visible tail, partly because the tail always points away from the Sun and at great distances therefore points away from the observer on Earth as well, and partly because the tail develops as the comet gets nearer the Sun. A comet also becomes brighter as it nears the Earth and the Sun.

In the case of a comet whose orbit is known, as mentioned above, the best course is to set out the orbit on a star map from published details of Right Ascension and declination. Then establish the position for the date of proposed observation and search the sky with a low-powered wide-angle eyepiece in the area indicated. Sweeping should be carried out in an orderly manner, for merely transversing the instrument erratically will most likely result in disappointment. Start at some point or star, and sweep across the estimated position of the comet, then depress or elevate the instrument for a space of half the diameter of the field of view, and pass over the area again. In this way an area of the sky will be covered completely and, if all is well, the comet should be found.

Comets are not usually the bright dramatic objects seen in pictures (Plate 19). Most of them are decidedly faint objects when first sighted, and interference in your observations by artificial light nearby can be a considerable nuisance. In urban areas it is sometimes possible to site the instrument in the deep shadow cast by a house or other object. This will not dispel the sky brightness but will assist in keeping the eyes dark-adapted. Obviously one does not use a bright light for reading the star maps or one's notes, though a dim red light is helpful.

Having detected a comet, you must next estimate its brightness and its angular dimensions, as well as its position. Brightness in terms of magnitude can be assessed by comparing the comet with stars of known magnitudes in the same manner as is employed for variable stars (page oo). One great difficulty is comparing the apparent magnitude of a comet, which is an extended object, with a star, which is a point source of light. The usual method is to put the stars out of focus until they become similar to the comet. Here trouble is encountered because the further a star is put out of focus, the fainter it becomes. The correct amount of defocusing is a matter of personal decision, and this alone tends to produce inaccuracies. At best, therefore, a comet's magnitude is a matter of estimation (see Appendix 3).

An alternative is to take a photograph, keeping the telescope centred on the comet and allowing the stars to drift. You can then estimate magnitude by examination of the picture, or, if possible, measurement of the density of the image.

A warning of the danger of seeing ghosts, or reflections from the lenses in the eypiece, is necessary. A ghost can be recognised if it disappears when you change the eyepiece, or sometimes you can detect them by merely moving the image to a different part of the field of view.

19 Comet Burnham, 1960, photographed at Dr Waterfield's observatory, Ascot, Surrey, England, with 6in (152mm) lens and exposure of 20 min. The telescope was made to follow the comet, and the stars appear as streaks (*H. B. Ridley and M. J. Hendrie*)

The size of the comet under observation can be ascertained by reference to the diameter of the field of view in the telescope. We know the Earth rotates once in 24 hr, so that an equatorial star will traverse the field, if the telescope is kept still, at the rate of 1° every 4 min. Time the transit of a star across the diameter of the field of view in the eyepiece, and from this work out the diameter of the field—4 min per degree, 2 min for half a degree, 1 min for a quarter of a degree, and so on. Now you can compare the apparent dimension of a comet with the diameter of the field, and convert to fractions of a degree.

The length of tail is a difficult matter, because the larger the telescope the more light it acquires, and so the longer the tail appears to be. Sky brightness also enters into the estimate. It is best, therefore, to record both the size of the instrument and the state of the sky when estimating tail lengths. Binoculars are sometimes useful for estimating tails of large comets because of the extended field of view.

While dealing with the tail of a comet, it is worth noting that the direction of the tail is no indication of the direction of travel of a comet. The tail always points away from the Sun, irrespective of the orbital motion. The reasons for this are outside the terms of reference of this book, but can be found in any good book on comets.

Under higher magnification the structure of a comet can be observed. Its head comprises a coma and sometimes a nucleus, and drawings of details of these (the nucleus is rarely seen) can be of value. The appearance of the head often changes rapidly. The behaviour of the tail should also be recorded, for straight tails indicate comets of a gaseous nature and curved tails indicate dusty ones.

To be of value, observations of comets should be properly recorded, as follows:

1 Enter the name, year and assigned letter (if available) of the comet.
2 The time of observation should be expressed thus: year, month, day and decimals of a day. Thus an observation made at 23 hr 30 min UT on 30 July 1973 should be expressed as 1973 July 30·98.
3 Approximate Right Ascension and declination should be given, and the epoch stated. (The epoch is the date to which the Right Ascension and declination are worked.)
4 Observing conditions should be listed, especially sky transparency—shown by stating the magnitudes of the faintest stars visible to the naked eye at the zenith and at the same altitude as the comet.
5 Give the magnitude estimate and comparison stars used for the observation.
6 (a) Describe the coma, giving its apparent diameter in minutes of arc, and say whether it is diffused, has a strong concentration at the centre,

has well marked edges, or any other features.

(b) Describe the nucleus, but do not confuse a small central condensation with a nucleus.

(c) Describe the tail (if any)—whether straight or curved, what its length and position angle are (measured from North through East in degrees), etc.

7 Detail the instrument used, with size and magnification.

8 Give name and location of observer.

9 Supply a field sketch identifying at least one star and showing the position of the comet. Also draw the appearance of the comet with special reference to the shape of the tail and its position angle and length. A good photograph is even better (Plate 19).

If you are fortunate enough to discover a new comet, or to recover one unobserved since its last appearance, advise your organisation immediately; in Great Britain this is the Comet Section of the British Astronomical Association. National observatories are also interested.

Observing the Stars

So far, observation has been related to individual bodies and with some considerable detail, but now we come to a different sphere of activity. The stars and nebulae are roughly millions of times further away from Earth than are the bodies so far considered, and all our knowledge has to be derived from their light. Stars only appear as points of light, any apparent disk originating in the optical system of a telescope or camera. This situation hampers observations considerably, but there are still many fields of activity open to users of reasonably sized instruments, although most of the research work is today undertaken by users of large telescopes and radiotelescopes in our professional observatories.

Until the advent of the space vehicle our knowledge of the bodies in the Solar System was to a reasonable extent the result of amateur observation. In the case of stars and nebulae the situation is different. Nevertheless, there are still fields where the user of a modest instrument can augment the work done at large observatories.

We shall consider these fields under three headings: binary stars, variable stars and observing with the spectroscope.

Binary Stars

The term binary stars usually covers pairs of stars held in orbit around each other by gravitational force, though there are many multiple systems, with more than two components, in which similar conditions prevail. Double stars may be considered a wider definition including stars that appear close merely because they are nearly in line of sight from the observer. We are not concerned with these.

For the observation of widely spaced binary stars a magnification of about 100 to 160 times may be sufficient, but for closely placed binaries something in the region of 200 to 400 times is necessary, according to the closeness of the two stars under observation. This means that small instruments are not satisfactory, and the observer is limited in many cases by the resolving power of his instrument. The relative resolving power of telescopes of various sizes is set out on page 10, but may be calculated for any diameter of glass by the formula $R = 5/D$ in inches, $R = 125/D$ in millimetres, where R is resolving power and D diameter. Thus a 10in (250mm) diameter objective or mirror should separate two stars at $0·5''$ of arc displacement. For general observation of binary stars the minimum size of telescope is about 8·5in (215mm).

A micrometer is essential for measuring the distance between the two stars in a binary system in seconds of arc; this has been described on page 26. The procedure is to close the wires, read the indicator, open the wires to accommodate the two stars accurately, and read the indicator again. Subtraction of the two readings will give the separation in terms of divisions of the scale, which are then converted to seconds of arc from values previously determined by reference to measurements of objects whose separation is already known—usually long period binary systems or planetary diameters, whose dimensions are published regularly. This procedure is repeated several times and a mean of the various values calculated.

It must be emphasised that once a micrometer has been calibrated on the sky in a certain telescope, the relation between the scale and actual seconds of arc on the sky only applies when the micrometer is used in that particular instrument. Transfer of the micrometer to another instrument or the substitution of another eyepiece will make recalibration necessary.

Having measured the separation of the binary components you must now find the position angle. This is measured from the bright star to the faint companion, on a circle with zero degrees at the North point and increasing values through East to South and West. A position-angle circle is usually fitted to a micrometer, and should be preset to read zero at the north point. This can be found by allowing the stars to drift across the field of view and establishing the east-west line, setting east at 90° and west at 270°.

When a large number of observations of a particular binary system has been obtained, sometimes needing many years of work, the data can be plotted on paper to show the apparent distances at various position angles. This will yield a picture of the apparent orbit

Generally the plane of the orbit is inclined to the line of sight, and only a projection is shown by the plot. This is usually an ellipse, but it must be remembered that the system revolves around the common centre of gravity of the two stars.

The larger star will not be at the focus of the ellipse, and the major and minor axes will not be at right-angles in the projected orbit. However, since equal areas project as equal areas, the less massive star will describe equal areas about the more massive one in equal times in the projected orbit.

Various methods are employed for transferring the projected orbit to a true orbit, but they are beyond the scope of this book. Reference may be made to standard works on stellar dynamics.

It will be recalled that large stars use up their energy more quickly than smaller ones. If binary systems are made up of stars of the same age, the difference in colour of the individual stars today will indicate the difference in their respective development. The colour combinations in binaries are often striking. Some notable cases are listed below:

Eta Cassiopeiae	Yellow and purple
Gamma Andromedae	Yellow and blue
Phi Tauri	Red and bluish
Rho Orionis	Yellow and blue
Alpha Scorpii	Red and emerald green
Epsilon Boötis	Yellow and blue green
Alpha Herculis	Yellow and emerald
Gamma Leonis	Yellow and greenish
Beta Cygni	Gold and blue
Zeta Sagittae	Light green and blue
Epsilon Draconis	Yellow and blue
Beta Cephei	Light green and blue
Sigma Cassiopeiae	White and blue

The list of representative binary stars in Table 13 may be helpful to the would-be observer. The position angles and distances are quoted for 1977.

Table 13

Some Binary Stars

Star name	RA hr	min	dec °	′	Magnitudes		Position angle	Distance (sec)	Period (years)
Eta Cas	0	46	N57	33	3·7	7·4	305·3	11·83	480
Alpha Gem	7	31	N32	00	2·0	2·9	104·5	2·05	420
Gamma Leo	10	17	N20	06	2·6	3·9	123·2	4·29	620
Xi Uma	11	15	N31	49	4·4	4·9	110·3	3·01	60
Zeta Boo	14	39	N13	57	4·6	4·6	306·2	1·12	125
Gamma Lup	15	32	S41	00	3·6	3·7	277·5	0·62	—
Zeta Herc	16	39	N31	41	3·1	5·6	162·6	1·18	34
Epsilon 1 Lyr	18	43	N39	37	5·1	6·1	355·9	2·70	1,200
Epsilon 2 Lyr	18	43	N39	37	5·1	5·4	84·9	2·33	600
Delta Cyg	19	43	N45	00	3·0	6·5	234·6	2·32	—
Zeta Aqu	22	26	S00	17	4·4	4·6	234·9	1·72	—

Variable Stars

Variable stars are those that vary in brightness. They were studied originally to establish the theory behind their variability, and to maintain records to check on mutations of the characteristic curves of their variation. The idea was to be able to fit them into the general scheme of stellar evolution. Today it is necessary to maintain the records to study the permissive diversions from normal behaviour. Continuity of observation is essential, otherwise false deductions can easily be made from inadequate data, so leading to inaccurate theory.

Variable stars fall into two main classes: extrinsic and intrinsic. In the former the variability arises from causes outside the body of the star. In this class we place such stars as Algol, which is an eclipsing binary system, although a third dark body is also present; in this case the smaller star is the brighter and the larger one the fainter. Beta Lyrae, another member of this class, suffers tidal distortion between its members.

Intrinsic variables are those whose variability arises from the star's own instability. The Cepheid variables are representative, as are the long-period old-type stars. Cepheid variables change in brightness in a regular fashion according to their masses, and they have periods of variation ranging from about 3 hr to 45 days. RR Lyrae stars are similar but smaller, and pulsate at a faster rate than the Cepheids. Another class of intrinsic variable stars is the eruptive type, with sudden outbursts of brightness. This class includes such stars as U Geminorum, T. Tauri, U Orionis and R Coronae Borealis, in addition to novae and supernovae.

It will be appreciated from the foregoing that the observation of variable stars is a matter of measuring in some way the apparent brightness of a

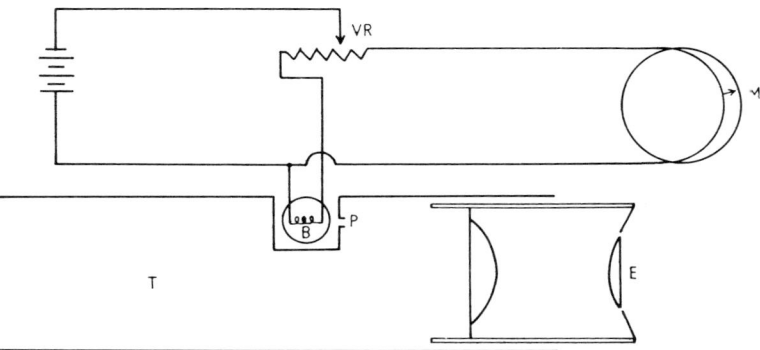

Fig 18 Visual photometer. VR, variable resistance; B, pea bulb; P, pinhole in screen enclosing pea bulb; T, telescope tube; E, eyepiece in which the image of P is produced to form an artificial star of controllable brightness; M, milliameter.

star at some particular instant. Therefore, for visual observation, one chooses stars with large variations in brightness and those within the telescope's light-collecting capability.

From the formula $M = 9 + 5 \log D$, where M is the magnitude of the faintest star visible in the telescope and D is the diameter of that instrument in inches, one can calculate the size of telescope required to reach a limiting magnitude of stars visible through it. Conversely, given the size of the telescope, one can calculate the faintest star visible through it. The resulting figures are shown in Table 1 (page 10).

It is thus useless to expect to see stars as faint as, say, magnitude 14 in a telescope of only 6in (150mm) diameter. Table 1 shows that an instrument of 15in (380mm) will be necessary to do this. The observer is therefore restricted in his choice of variable stars to those that do not drop below the limiting magnitude his instrument can reach.

As to the selection of stars with large light fluctuations, and hence suitable for visual observation, this will include long-period variables with periods of variation ranging from 2 months to 2 years, irregular variables such as SS Cygni, and some of the flare stars that remain faint for most of their lives but suddenly, and often unexpectedly, flare up and become temporarily brighter than normal. Novae too are frequently to be noted by the alert observer.

Serious observers who contact the appropriate organisations will gladly be given guidance on this matter, and will be recommended certain stars within the capabilities of their instruments. Then, having made the choice and been supplied with appropriate charts of the starfields to be used for finding the variable star, the observer will next have to find the variable itself.

First, estimate the position roughly by reference to a star atlas. Second, mark on the star chart the bright stars near the variable and draw in a series of lines connecting them; this produces a figure that the mind can easily remember. Third, with the telescope and a low power of magnification for the sake of having a large field, work systematically through the design drawn on the chart and gradually find the way to the variable itself. Having done this, confirm, by tracing the outlying brighter stars, that you have correctly recognised the field.

Now comes the problem of estimating the apparent magnitude of the variable star (see Appendix 4). At this stage one immediately thinks of some form of light-measuring instrument such as the Zollner photometer. (See p. 27, on which some of the difficulties encountered in the operation of photometers were discussed.) I have done some experiments over the years with a photometer employing a variable artificial star, which is adjusted to equal the brightness of the variable star, the light level then being read off on a milliameter scale (see Fig 18). Numerous difficulties will be encountered; the main ones are detailed below.

In the first place, the reading on the milliameter scale will initially be nowhere near the true apparent magnitude of the star observed. The photometer will have to be calibrated against actual stars by direct observation. Then the scale will most likely be of a different gradation from the stellar magnitude scale, which will entail calibration against a large number of stars whose magnitudes are accurately known (this information can be obtained from a star catalogue), through the whole range of the milliameter scale. It is not good enough merely to select a few stars and hope the rest of the scale will follow the same curve; the whole scale must be checked.

Sky clarity, which varies from night to night, will always have to be taken into account. The photometer will have to be recalibrated to find the proper constant to be applied to the readings taken that night.

Atmospheric absorption will also have to be taken into account, for it is always present in some degree and has quite disastrous effects if it is ignored. The sky may be comparatively clear near the zenith, but nearer the horizon it will be seen to be progressively more contaminated by mist and perhaps light cloud. The photometer will need recalibrating for use at various altitudes.

Finally, we come to a more subtle effect. When the lightbulb in the photometer is at its brightest, its light approaches white in colour. When reduced current is applied, the lower level of light contains an admixture of red light. To some observers, myself included, red light seems brighter than white or blue light of equal intensity. This throws the observations out of true, and means that magnitude readings for faint stars will suggest that they are brighter than they really are.

Similar difficulties bedevil all forms of photometer in one way or another, and a system that will circumvent them is altogether more suitable. The system used by many visual observers of variable stars is that in which the magnitude of the variable is directly compared with those of other stars whose magnitudes are accurately known, using a low-power eyepiece with large field (a higher power will be found useful when the variable is very faint). This means that the variable is compared with other stars in the same field, and therefore at the same altitude, so that any atmospheric effects are common to both the variable and the comparison stars.

There are two standard methods of accomplishing this. The first to be considered is the *Fractional Method*.

In this the variable is compared with two other stars, one slightly brighter and the other slightly fainter than the variable. The difference should not be more than about half a magnitude, if possible, to avoid errors due to the different interval of brightness as one moves down the magnitude scale—the ratio of difference between two magnitudes (by the factor of 2·512) means the magnitude scale is governed by a geometric progression, so that the difference in light level between stars of, say, magnitudes 4 and 5 is less than that between stars of magnitudes 10 and 11.

The interval of light intensity between the two comparison stars is imagined to be on a scale of, say, five divisions. The variable is mentally placed on this scale, and may give an impression of being, say, one step fainter than star A and four steps brighter than star B. This would be written A(1) V(4) B. The bright star is always quoted first for the sake of uniformity and to assist the coordinator of observations. The great advantage of this sytem is that both ends of the light scale are firmly fixed.

The next step is to deduce from this proportion the exact magnitude of the variable, as follows:

Take the recorded estimate A (1) V (4) B (noting that the proportions are written within brackets for the sake of clarity) and write down the magnitudes of the comparison stars. Suppose that Star B is magnitude 6·43, and Star A magnitude 5·84, figures obtainable from the catalogue or the sequence supplied by the appropriate organisation. The difference between the magnitudes is 0·59. Having used five divisions of the scale, we must now find the value of one division, so we divide the difference by five, and get 0·118. Four-fifths of the difference is 0·472.

Now return to the original estimate A (1) V (4) B and note that we want one-fifth added to the magnitude of Star A and four-fifths subtracted from the magnitude of Star B to fix the magnitude of the variable, thus:

Star A	mag 5·84	Star B	mag 6·43
Add one-fifth	0·118	Subtract four-fifths	0·472
	5·958		5·958

Having checked that the figures agree, we can record the magnitude of the variable as mag 5·958. In point of fact the human eye cannot discriminate smaller fractions than one-tenth of a magnitude, so we round the figure off to magnitude 6; i.e., to the nearest tenth of a magnitude.

It sometimes happens that the variable is brighter than any comparison star on the chart and sequence list. In this case we might record the estimate as V (2) A (3) B. V, being the brightest star, is quoted first, and represents the variable as brighter than the two comparison stars. We take the difference between the comparison stars' magnitudes in the same way as before. Suppose that Star B is magnitude 7·6 and Star A is magnitude 7·0. The difference is then 0·6.

Here we must be careful. The difference in magnitudes is only three-fifths of the whole five-step scale. To complete the series we require another two-fifths for the interval V (2) A. If three-fifths of the interval is 0·6, then two-fifths is 0·4.

Now take the original estimate V (2) A (3) B and substitute magnitude figures for the letters. Then,

$$V \ (0·4) \quad 7 \ (0·6) \quad 7·6$$

From this expression it will be seen that V is 7 minus 0·4, or 6·6. At the same time the expression can be checked by taking the variable's magnitude of 6·6 and subtracting the whole five-fifths of the scale V to B. Thus

Star B is	7·6
Variable is	6·6
Difference is	1·0

which is the same as the difference between Star B and the variable over the whole of the scale of five-fifths.

The second method, known as *Pogson's Step Method*, recognises that the human eye can estimate light to an accuracy of one-tenth of a magnitude. The variable star is compared with a number of comparison stars and estimates of the differences are made in steps of about one-tenth of a magnitude. The actual step interval is particular to each individual observer in practice.

The estimates are written down C−2, D+3 etc., meaning that the variable star is two steps fainter than star C and three steps brighter than star D. A series of such estimates is made, and, after reduction to their respective magnitudes, an average is calculated to represent the true magnitude of the variable.

This method can yield accurate results, but it has its disadvantages also. The system requires eye training and may be found difficult by a beginner. The length of the individual steps varies with fatigue and the health of the observer, so that the moral is: Do not observe when you are tired!

In either method it is necessary to get a good straight-ahead view of the stars, both variable and comparisons. This will necessitate bringing the stars to the centre of the telescope field—equidistant from the centre in the case of close comparison stars. When comparison stars are situated some distance from the variable, swinging the telescope to get them into view can cause errors. The brain is quite untrustworthy at remembering star magnitudes.

Remember also that the eye is somewhat untrustworthy. The retina is more sensitive at the bottom than at the top through daytime fatigue; also, a star nearer the nose will appear brighter than one at the outer side of the face, and a star at the edge of the field brighter than one at the centre. The presence of bright stars in the field of view also upsets one's judgement of the brightness of fainter stars. Most variable stars are ruddy in colour and have to be compared with blue normal stars.

To avoid spoiling one's night dark-adaptation of the eyes it is suggested that a dim red light be used to read with, and that variable stars are located by reference to star charts rather than by using setting circles. Reading the circles can destroy the sensitivity of the eye under dark conditions.

Observations should be classified as follows:

Class 1   a satisfactory and trustworthy observation;
Class 2   observation made in hazy or cloudy conditions and not as trustworthy as class 1; and
Class 3   observations made under difficult conditions, such as cloud or artificial light interference, and not considered reliable.

Classes 1 and 2 observations may be recorded for use, but class 3 observations are usually rejected.

A record should be kept for each star in a ledger, one page to each star. In this way rogue observations will be spotted by inspection of the magnitude column. When plotted on a graph, these magnitudes (if enough have been obtained) will produce a curve from which the period of variation of the star can be deduced. This observed curve can be compared with the characteristic curve of the magnitudes to discover any undue variation in the star's behaviour.

Variable star statistics are referred to Julian Days, not normal civil

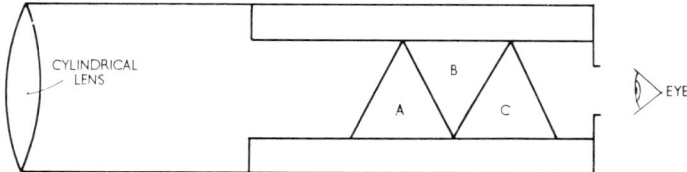

Fig 19 Section diagram of Zollner, or direct vision, spectroscope. A, B and C are three prisms.

dates, for the sake of regular continuity over long periods. While the observations are recorded against date and time, it will be necessary to work out also the Julian Day and the time expressed as a decimal of a Julian Day, and to plot curves to this dimension.

It is suggested that the ledger entries be made under the following headings:

Name of Star, and location of observer.
Instrument and magnification used.
Year; Date; time (GMAT beginning at noon).
Julian day and decimal.
Light estimate.
Deduced magnitude.
Class of observation.
Remarks—such as hazy, cloud, twilight, etc.

GMAT is Greenwich Mean Astronomical Time and is 12 hr slow on UT. It is used to avoid the change of date during the night.

Observing with the Spectroscope

The simplest form of spectroscope is the Zollner (or direct vision) type, with its three prisms in line, as in Fig 19. This instrument will show the

Fig 20 Layout of a typical star spectroscope

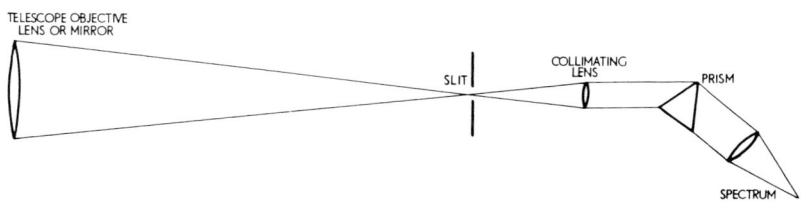

different sorts of stellar spectra, but is too small for serious study.

The more sophisticated type is equipped with one or more prisms, set at suitable angles for the light to pass through each of them in succession, as in Fig 20. This gives greater dispersion.

Spectroscopy is an involved science which should be studied with reference to specialist books, one of which is mentioned in the Bibliography (page 110).

One must admit that the owner of a small or medium-sized telescope can do little that is not already better done in professional observatories, but it is interesting to see and appreciate for oneself the wonders of the Universe as displayed by the spectroscope.

Observing Star Clusters, Nebulae and Galaxies

In this area the amateur is heavily outclassed by the professional. But there are many objects of interest worthy of inspection by the owner of a medium-sized instrument.

One field of activity is the search for novae and supernovae, which occur at magnitudes ranging from 8 to 10 at their brightest, in galaxies and clusters. Long hours of patrolling, and an intimate knowledge of the appearances of the galaxies and clusters, are necessary for this study; an instrument of 14in (350mm) diameter or larger is recommended. Those interested in celestial photography can do valuable work: photographs taken at different dates can be compared in the hope of recording a new star or a variable which has changed in brightness during the interval between the taking of the photographs.

On the less technical side, satisfying views of many beautiful nebulae, galaxies and star fields can be obtained with a low-power eyepiece to give a large field of view. Some of these fields lend themselves to photography and a specimen is shown in Plate 20.

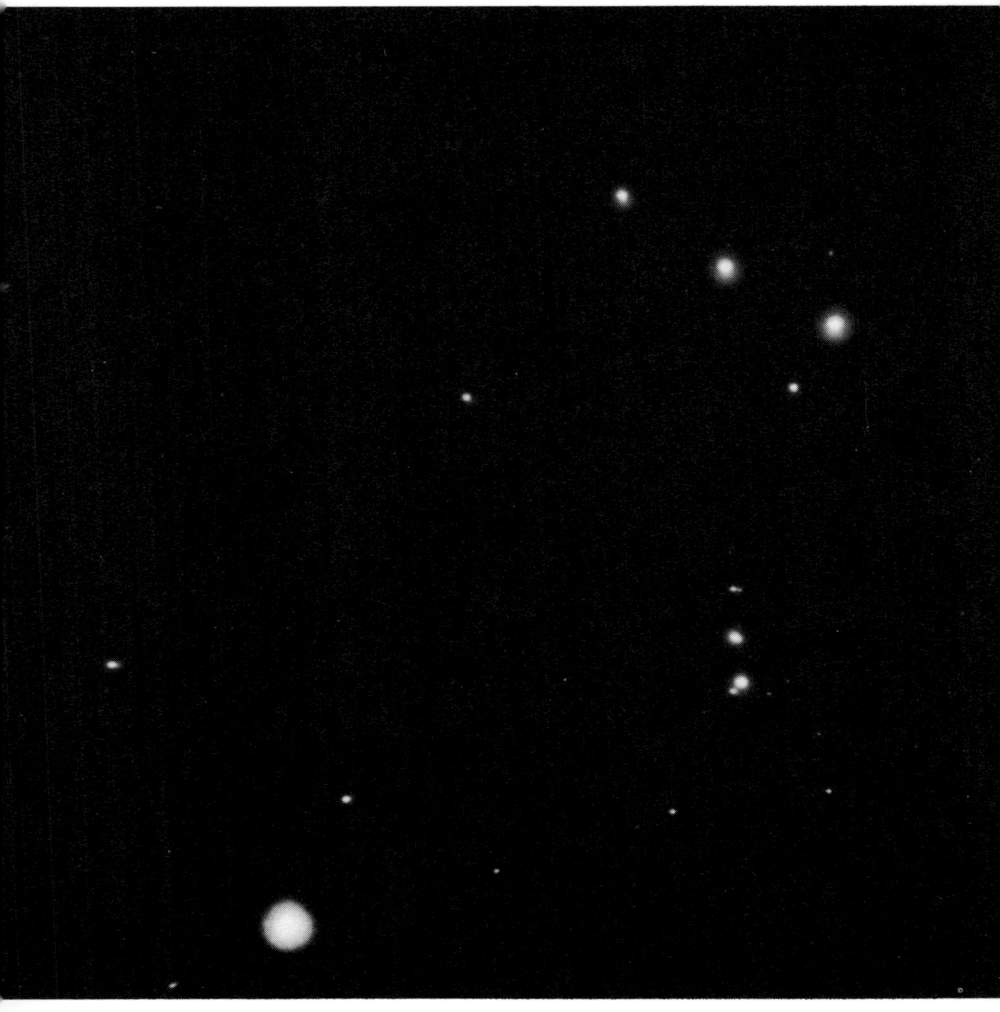

20 Southern portion of constellation of Orion, showing the Orion Nebula. Photograph by the author 1973 February 24, with 3in (76mm) lens at f/3 and exposure of 10 min.

Lists of many objects are available in such publications as *Norton's Star Atlas* and Webb's *Celestial Objects for Common Telescopes*, while the items in Messier's Catalogue are worthy of inspection.

Brief Glossary

Celestial Sphere	An imaginary sphere, centred on the Earth, on which, for the purposes of angular computation, celestial bodies are considered to be situated.
Cusps	The points of a crescent.
Declination	Angular distance north or south of the celestial equator.
Dichotomy	Exact half phase of the Moon or a planet.
Elongation	Apparent angular distance of a body from the Sun, or of a satellite from its primary body.
Ephemeris	Table showing the predicted motions of a moving celestial body.
Equator, celestial	The projection onto the celestial sphere of the Earth's equator.
Galaxy	A separate star system such as our own Galaxy (the Milky Way) or the Great Andromeda Nebula, M31.
Limb	The edge of the disk of a planet or other celestial body as seen from Earth.
Magnitude	*Apparent magnitude* is the brightness of a celestial body as seen from Earth. *Absolute magnitude* is the apparent magnitude the celestial body would have were it to be placed at a distance of 10 parsecs (about 33 light years) from Earth.
Nebula	Mass of tenuous gas and or dust in space. Also used for distant galaxies but, owing to possible confusion, this practice is dying out.
Opposition	Position of a body exactly opposite the Sun in the sky; i.e., so that the Earth lies on a straight line joining the celestial body and the Sun.
Poles	North and south points defined by the rotation of the Earth or other celestial body. The *celestial poles* are the projection onto the celestial sphere of the terrestrial poles.
Reflector	Telescope working by light reflected from a mirror.
Refractor	Telescope working by light refracted through a lens.
Right Ascension	Angular distance, measured along the celestial equator, from the First Point in Aries.
Sidereal	Relating to a star, as in *sidereal period*.
Synodic	Relating to the Earth, as in *synodic period*.
Terminator	Boundary between day and night on the surface of a planet or satellite.
Zenith	The point on the celestial sphere directly overhead.

Appendix 1

Magnitudes of Asteroids

The mean opposition magnitudes of asteroids are tabulated from many observations. This tabulation includes a constant g from the equation

$$g = m - 5 \log a (a - 1)$$

where a is the asteroid's mean distance from the Sun in astronomical units and m is the mean opposition magnitude. If one knows the constant g, one may calculate the actual magnitude at any particular time from the equation

$$m = g + 5 \log rD$$

where r is the distance of the asteroid from the Sun in astronomical units, D is its distance from Earth, and g is the constant derived as above.

Individual asteroids may differ from the derived figures because they have different albedos.

In addition to the above, the phase angle q must often be considered, especially with the larger asteroids. This angle, the angle Earth-asteroid-Sun, is given by the following equation:

$$\cos q = \frac{r^2 + D^2 - R^2}{2rD}$$

where r is the distance from the Sun to the asteroid, R is that of the Earth, and D is the distance of the asteroid from Earth. (Working in astronomical units, this equation becomes

$$\cos q = \frac{r^2 + D^2}{2rD} \bigg).$$

The phase, k, is calculated from

$$k = \tfrac{1}{2} (1 + \cos q).$$

The phase angle can cause a variation of magnitude from 0·5 to 1·7. (See J. L. White in *Practical Amateur Astronomy*, edited by Patrick Moore, pages 136 *et seq.*)

Appendix 2

Longitude on Saturn

Although no datum point for longitude on Saturn has been officially fixed, the American Association of Lunar and Planetary Observers, together with the Saturn Section of the British Astronomical Association, have adopted two systems similar to those used on Jupiter. System I applies to the South Equatorial belt, the Equatorial Zone and the North Equatorial belt, and is based on a rotation period of

10 hr 14 mins. This system showed the zero point at the central meridian of Saturn's disk on 1977 November 23 at 23 hr 7 min 40 sec UT. System II applies to the tropical and temperate zones as well as the polar regions. This system is based on a rotation period of 10 hr 38·5 min, and showed the zero point on the central meridian of Saturn's disk on 1977 November 29 at 23 hr 16 min 30 sec UT.

Copies of the tables are available to members of either of the organisations mentioned above. They are also published in *The Strolling Astronomer*.

Appendix 3

Magnitudes of Comets

The apparent brightness of a comet varies inversely with its distance from the Sun and Earth according to certain principles, though no comet behaves accurately in this way. We accordingly adopt general expressions in terms of stellar magnitudes for each of the main classes of comets: for comets with parabolic orbits it is $M = m + 10 \log r + 5 \log D$, and for short-period comets it is $M = m + 15 \log r + 5 \log D$, where M is the observed magnitude of the comet, m is the magnitude it would have if the distance from both Sun and Earth were unity (i.e., 1 astronomical unit), r is the actual distance from the Sun and D is the actual distance from Earth.

The different values of m observed for different comets are attributable to their physical and chemical compositions. Fluctuations also occur when a comet is a long way from the Sun and the temperature is too low to expel gaseous matter from the nucleus.

A third equation, $M = 5 + 5 \log D + 10 \log r$, is often used to give a magnitude derived from the proportionate distances of a comet from Earth and Sun.

Appendix 4

Stellar Magnitudes

The classification of stars according to their apparent brightness was first carried out by Hipparchus at Rhodes and Alexandria. He made a catalogue of 1,080 stars between 130 and 127 BC, grouping all the brightest stars in the first magnitude and the faintest stars in the sixth. Whether he had a definite idea of light ratio in making the division or whether it was mere accident that he chose six classes is not known. Ptolemy, who followed Hipparchus, does not say, but the choice is significant. Ptolemy himself, in AD 140, issued his *Almagest* in thirteen volumes; he assigned magnitudes to the stars, and introduced subdivisions by using the words 'greater' or 'less'.

In 1852-63 Argelander, with Schönfeld and Krüger, compiled the *Bonn Dürchmusterung Catalogue*, using decimal divisions of stellar magnitudes, and about this time Sir John Herschel noted that a decrease of light in geometric progression corresponded to an increase of magnitude in arithmetic progression. This agreed with Fechner's Law, which states that as a stimulus increases in geometric progression, its resulting sensation increases in an arithmetic progression, but there is a slight deviation when the stimulus becomes very intense or very slight. The relation may be expressed as $S=C \log R$, where S is the intensity, R the stimulus and C is a constant.

As we have no absolute standard of brightness we can only compare two brightnesses. Thus, if we let A and B represent the brightnesses of two stars, and M_1 and M_2 their magnitudes, $M_1 = C \log A$ and $M_2 = C \log B$.

It follows that if star A is 100 times brighter than star B the difference in magnitude is $2C$, since log 100 is 2. It was also established that a star of the first magnitude is just about 100 times as bright as a star of the sixth magnitude.

Accordingly, in 1850, Pogson proposed that a fixed scale be adopted as the uniform ratio between magnitudes. For ease in calculation he suggested the factor 2·512, whose logarithm to the base 10 is 0·4. The relative magnitude differences and the brightness ratios are set out in Table 14.

Table 14

Magnitudes and Brightness Ratios

Magnitude difference	Brightness ratio
0·1	1·096
0·2	1·202
0·3	1·318
0·4	1·445
0·5	1·585
0·6	1·738
0·7	1·905
0·8	2·089
0·9	2·291
1·0	2·512
2·0	6·310
3·0	15·849
4·0	39·811
5·0	100·000

If we take the number 2·512 and multiply it by itself (to the third place of decimals), we have the following series: 2·512, 6·310, 15·851, 39·818, 100·023. For all practical purposes these figures agree with those for whole-number magnitude differences in Table 14. The magnitude scale is thus not really a scale of brightness, but rather a scale of faintness because the figures grow larger as the brightness grows less. The mathematical derivation of the theory is well outlined in *An Introduction to the Study of Variable Stars* by Caroline Furness and in *Essentials of Astronomy* by Motz and Duveen (see Bibliography).

Table 15 gives a list of standard stars for magnitude comparison.

Table 15

Magnitudes of Standard Stars

Approximate magnitude	Name of star	Exact magnitude
1·5	Alpha Geminorum	1·58
	Lambda Scorpii	1·60
	Gamma Orionis	1·64
2	Beta Ursa Minoris	2·04
	Kappa Orionis	2·06
	Alpha Andromedae	2·06
2·5	Gamma Ursae Majoris	2·44
	Epsilon Cygni	2·46
	Alpha Pegasi	2·50
	Delta Leonis	2·57
3	Zeta Aquilae	2·99
	Gamma Boötis	3·05
	Delta Draconis	3·06
	Zeta Tauri	3·07
3·5	Alpha Trianguli	3·45
	Zeta Leonis	3·46
	Beta Boötis	3·48
	Epsilon Tauri	3·54
4	Beta Aquilae	3·90
	Gamma Coronae Borealis	3·93
	Delta Ceti	4·04
	Delta Cancri	4·17
4·5	Nu Andromedae	4·42
	Delta Ursae Minoris	4·44
	Nu Cephei	4·46
	Psi Ursae Majoris	4·54
5	Rho Ursae Majoris	4·99
	Eta Ursae Minoris	5·04
	Delta Trianguli	5·07
	Zeta Canis Minoris	5·11
5·5	Theta Ursae Minoris	5·33
	Rho Coronae Borealis	5·43
	Epsilon Trianguli	5·44

Appendix 5

National Bodies

American Association of Variable Star Observers
187 Concord Avenue, Cambridge, Mass. 02138, USA.
Association of Lunar and Planetary Observers
American College Observatory, Edinburg, Texas, USA.

British Astronomical Association
Burlington House, Piccadilly, London W1V oNL, UK.
(The BAA has a New South Wales branch for the benefit of members in Australia, New Zealand, Tasmania and Fiji, etc.)
British Meteor Society
26 Adrian St, Dover, UK.
Junior Astronomical Society
58 Vaughan Gardens, Ilford, Essex, UK.
Royal Astronomical Society of Canada
252 College St, Toronto 2B, Ontario, Canada.

Bibliography

Furness, C. E. *An Introduction to the Study of Variable Stars* (Houghton Mifflin, Boston and New York, 1915)
Handbook of the British Astronomical Association (published annually)
Moore, Patrick (ed). *Astronomical Telescopes and Observatories for Amateurs* (David & Charles, 1973)
Moore, Patrick. *The Atlas of the Universe* (Mitchell Beazley, London, and Rand McNally, New York, 1970)
Moore, Patrick (ed). *Practical Amateur Astronomy* (Lutterworth Press, 1963)
Motz, L. and Duveen, A. *Essentials of Astronomy* (Blackie & Son Ltd, and Wadsworth Publishing Co Inc, Belmont, California, 1966)
Observer's Handbook, The (published annually by the Royal Astronomical Society of Canada)
Robinson, J. Hedley. *Astronomy Data Book* (David & Charles, 1972)
Sidgwick, J. B. *Amateur Astronomer's Handbook* (Faber & Faber, 1971)
Sidgwick, J. B. *Observational Astronomy for Amateurs* (Faber & Faber, 1971)
Thackeray, A. D. *Astronomical Spectroscopy* (Eyre & Spottiswoode, 1961)
Webb, T. W. (revised by Mayall, M. W.). *Celestial Objects for Common Telescopes*, two vols (Dover Publications, New York, 1962)

Index

Antoniadi Scale of seeing, 46
apodising screen, 28, 61
asteroids, 73, 106
atmospheric absorption, 27, 98

binary stars, 94, 96
binoculars, 9, 74, 88

clock, 32-6, 56
clothing, 7
colour filters, *see* filters
comets, 89, 93, 107

declination, 35, 89
definitions of terms, 105
diameter of field, 92
double stars, 94

eclipses:
    Jupiter's satellites, 79
    Moon, 56
    Sun, 46
eyepieces, 14-17

fading occultations, 56
filters, 27, 52, 61, 65, 67, 72, 79,
    85

galaxies, 103-4
grazing occultations, 56

limiting magnitudes, 10

magnification, 66, 69, 76, 80, 87
magnitudes, 90, 97-103, 107
maps:
    Mars, 71
    Moon, 48-51

Mars, 67-73
Mercury, 57-61
micrometers, 26-7, 94
Moon, 47-56,
    eclipses of, 56
    maps of, 48-51
    occultations by, 54, 56
    photography of, 21
    transient phenomena on, 47
mutual eclipses, etc., of Jupiter's
    satellites, 79

naked eye observation, 9
national bodies & societies, 109
nebulae, 103-4
Neptune, 88
novae, 103-4

observatories, 13
occultations:
    fading, 56
    grazing, 56
    of stars by the Moon, 54
    of stars by Saturn's rings, 87
occulting bar, 66, 85

photography, 19-25
    comets, 19, 90
    Moon, 21
    nebulae, 19
    planets, 21
    stars, 19
    Sun, 23
photometers, 27, 98
planets:
    asteroids, 73-5, 106
    Jupiter, 75-80
    Mars, 67-73

Mercury, 57-61
Neptune, 88
Pluto, 88
Saturn, 80-7, 106
Uranus, 87-8
Venus, 61-7
photography of, 21-23
Pluto, 88
position, 32-6
precession, 36

reading lamp, 101
resolution, 10, 94
right ascension, 33, 89

Saturn, 80-7, 106
latitude and longitude on, 81, 83, 106
seeing conditions, 46
spectroscope, 102-3
spectrohelioscope, 46
stars:
atlases of, 89
binary, 94, 96
clusters, 103
colours, 95
magnitudes of, 97, 107
novae, 103
occultations by Moon, 54, 56; by Saturn's rings, 87; fading, 56; grazing, 56
photography of, 19
supernovae, 103
variable, 96-102
Sun
eclipses of, 46
faculae, 38
filters, 37, 38

graticule, 39
latitude and longitude on, 40, 43
limb darkening, 38
photography of, 23
projection method of observing, 37, 38, 39
prominences, 46
solar eyepiece, 37
sunspots, 39
Wilson effect, 44
Zurich numbers, 43
supernovae, 103

Telescopes:
adjusting equatorial mounting, 17
cleaning, 29, 31
collimation, 30
dew cap, 28
eyepieces, 14-17
housing, 13
limiting magnitudes, 10
maintenance, 28
mountings, 11
reflector, 9, 29
refractor, 9, 28
resolution, 10
time, 32-6, 55, 56
transient phenomena on Moon, 47

Uranus, 87-8

Venus, 61-7

Wilson effect, 44

Zollner photometer, 27, 98
Zollner spectroscope, 102
Zurich numbers, 43